It's another Quality Book from CGP

This book is for anyone doing GCSE Geography.

Whatever subject you're doing it's the same old story — there are lots of facts and you've just got to learn them. KS4 Geography is no different.

Happily this CGP book gives you all that important information as clearly and concisely as possible.

It's also got some daft bits in to try and make the whole experience at least vaguely entertaining for you.

What CGP is all about

Our sole aim here at CGP is to produce the highest quality books — carefully written, immaculately presented and dangerously close to being funny.

Then we work our socks off to get them out to you — at the cheapest possible prices.

Contents

Section 1 — Tectonic Activity

Section 2 — Rocks and Resources

Section 3 — Weather and Climate

Section 4 — The Living World

Section 5 — Rivers

Section 6 — Glaciation

Section 7 — Coasts

Section 8 — Population

Section 9 — Urban Environments

Section 10 — Rural Environments

Section 11 — Development

Section 12 — Industry

Section 13 — Globalisation

Section 14 — Tourism

Exam Skills

Published by CGP

Editors:
Claire Boulter, Ellen Bowness, Joe Brazier, Karen Wells.

Contributors:
Rosalind Browning, Barbara Melbourne, Helen Nurton, Sophie Watkins, Dennis Watts.

Proofreading:
Glenn Rogers, Julie Wakeling.

Coordinated by Paddy Gannon

ISBN: 978 1 84762 429 1

With thanks to Laura Jakubowski for copyright research.

With thanks to iStockphoto.com for permission to reproduce the photographs used on pages 5, 19, 26, 31, 32, 34, 35, 36, 39, 53, 57, 69, 77, 81 and 104.

With thanks to Science Photo Library for permission to reproduce the photographs used on pages 9, 12, 20 and 86.

Images of Sri Lankan coastline on page 13 © UPPA/Photoshot.

Map of UK geology on page 15 reproduced by permission of the British Geological Survey. © NERC. All rights reserved. IPR/122-32CT.

Graphs of rainfall and sunshine hours on page 24 adapted from Crown Copyright data supplied by the Met Office.

Graph of the last 1000 years of climate change on page 27 adapted from IPCC, 2013: In: Climate Change 2013: The Physical Science Basis. Contribution of Working Group I to the Fifth Assessment Report of the Intergovernmental Panel on Climate Change, Box TS.5, Figure 1(b) [Stocker, T.F., D. Qin, G.-K. Plattner, M. Tignor, S.K. Allen, J. Boschung, A. Nauels, Y. Xia, V. Bex and P.M. Midgley (eds.)]. Cambridge University Press, Cambridge, United Kingdom and New York, NY, USA.

Graph of the last 150 years of climate change on pages 27 and 67 adapted from Crown Copyright data supplied by the Met Office.

Map of drought risk on page 34 © 'UCL Global Drought Monitor'.

Mapping data on pages 56, 70, 80, 167 and 168 reproduced by permission of Ordnance Survey® on behalf of HMSO © Crown copyright (2010). All rights reserved. Ordnance Survey® Licence No. 100034841.

Data used to compile the UK population density maps on pages 62 and 162 from Office for National Statistics: General Register Office for Scotland, Northern Ireland Statistics & Research Agency. © Crown copyright reproduced under the terms of the Click-Use Licence.

Data used to compile the UK average rainfall map on page 62 from the Manchester Metropolitan University.

Image of Rhône Glacier in 2008 on page 67 © Juerg Alean, Eglisau, Switzerland, http://www.glaciers-online.net

Data used to produce the Rhône Glacier graph on page 67 © VAW / ETH Zurich, http://www.glaciology.ethz.ch/swiss-glaciers/

Data used to construct the UK population pyramid on page 95 © Crown copyright reproduced under the terms of the Click-Use Licence.

World Population Graph on page 100 reproduced with kind permission from Jean-Paul Rodrigue (underlying data from the United Nations).

With thanks to Ellen Bowness for the desert and polar bear images on page 156.

With thanks to Mr Steve Ellingham for the image of trekking in the Himalayas on page 156.

Data used to construct the flow map of immigration on page 166 - Source International Passenger Survey, Office for National Statistics © Crown copyright reproduced under the terms of the Click-Use Licence.

Every effort has been made to locate copyright holders and obtain permission to reproduce sources. For those sources where it has been difficult to trace the copyright holder of the work, we would be grateful for information. If any copyright holder would like us to make an amendment to the acknowledgements, please notify us and we will gladly update the book at the next reprint. Thank you.

Printed by Elanders Ltd, Newcastle upon Tyne.
Clipart from Corel®

Based on the classic CGP style created by Richard Parsons.

Tectonic Plates

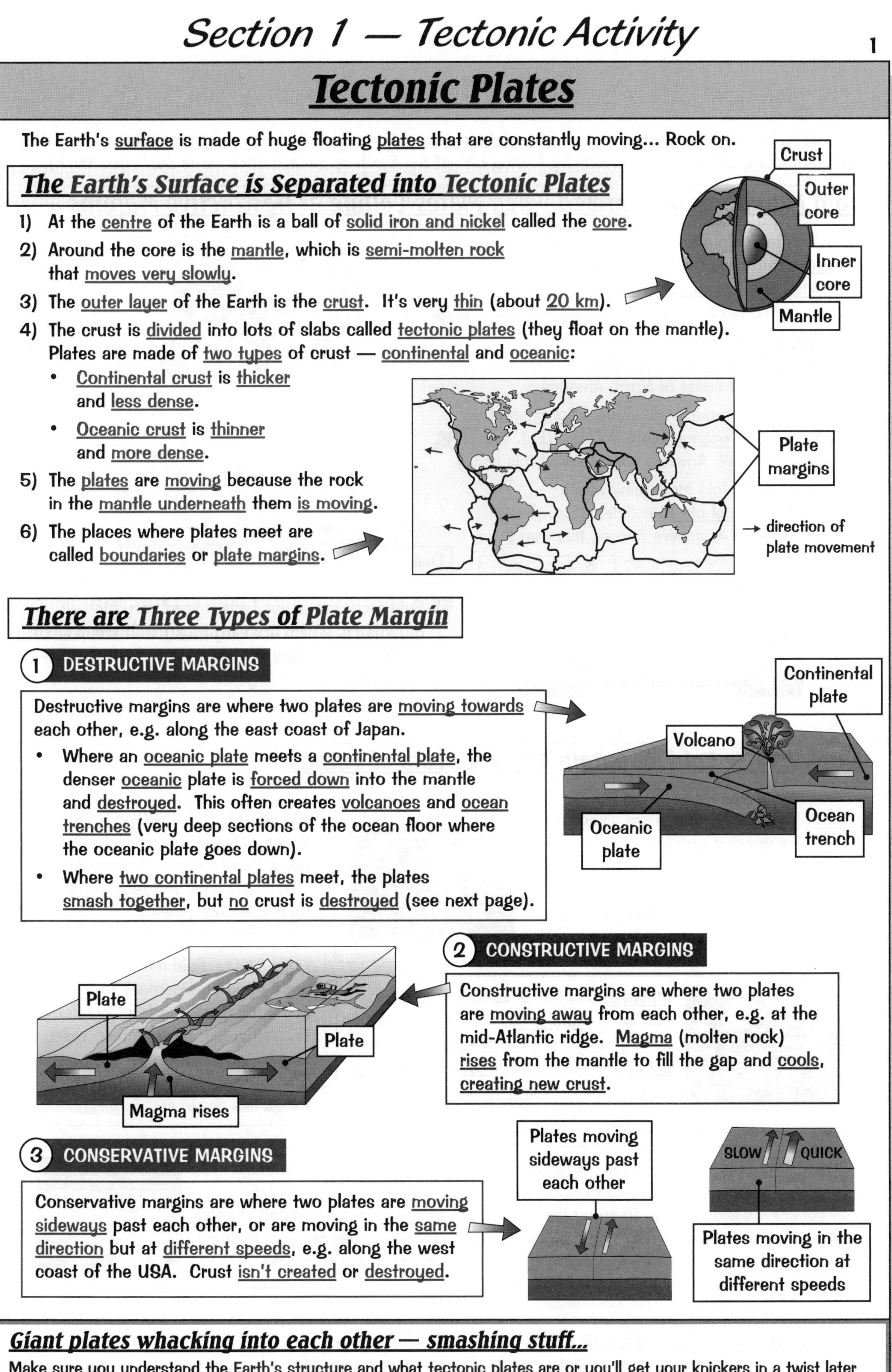

The Earth's surface is made of huge floating plates that are constantly moving... Rock on.

The Earth's Surface is Separated into Tectonic Plates

1) At the centre of the Earth is a ball of solid iron and nickel called the core.
2) Around the core is the mantle, which is semi-molten rock that moves very slowly.
3) The outer layer of the Earth is the crust. It's very thin (about 20 km).
4) The crust is divided into lots of slabs called tectonic plates (they float on the mantle). Plates are made of two types of crust — continental and oceanic:
   - Continental crust is thicker and less dense.
   - Oceanic crust is thinner and more dense.
5) The plates are moving because the rock in the mantle underneath them is moving.
6) The places where plates meet are called boundaries or plate margins.

There are Three Types of Plate Margin

1 DESTRUCTIVE MARGINS

Destructive margins are where two plates are moving towards each other, e.g. along the east coast of Japan.

- Where an oceanic plate meets a continental plate, the denser oceanic plate is forced down into the mantle and destroyed. This often creates volcanoes and ocean trenches (very deep sections of the ocean floor where the oceanic plate goes down).
- Where two continental plates meet, the plates smash together, but no crust is destroyed (see next page).

2 CONSTRUCTIVE MARGINS

Constructive margins are where two plates are moving away from each other, e.g. at the mid-Atlantic ridge. Magma (molten rock) rises from the mantle to fill the gap and cools, creating new crust.

3 CONSERVATIVE MARGINS

Conservative margins are where two plates are moving sideways past each other, or are moving in the same direction but at different speeds, e.g. along the west coast of the USA. Crust isn't created or destroyed.

Giant plates whacking into each other — smashing stuff...

Make sure you understand the Earth's structure and what tectonic plates are or you'll get your knickers in a twist later on in the section. Practise sketching and labelling the diagrams at the bottom to learn the types of margin too.

Fold Mountains

Get ready for the first landform created by plate movement. Drum roll please... fold mountains — ta dah. Fold mountains are mountains made by folding bits of the Earth up (imaginative name don't you think).

Fold Mountains are Formed when Plates Collide at Destructive Margins

1) When tectonic plates collide the sedimentary rocks that have built up between them are folded and forced upwards to form mountains.
2) So fold mountains are found at destructive plate margins and places where there used to be destructive margins, e.g. the west coast of North America.
3) You get fold mountains where a continental plate and an oceanic plate collide. (E.g. the Andes in South America were formed this way.)
4) You also get fold mountains where two continental plates collide. (E.g. the Himalayas in Asia were formed this way.)

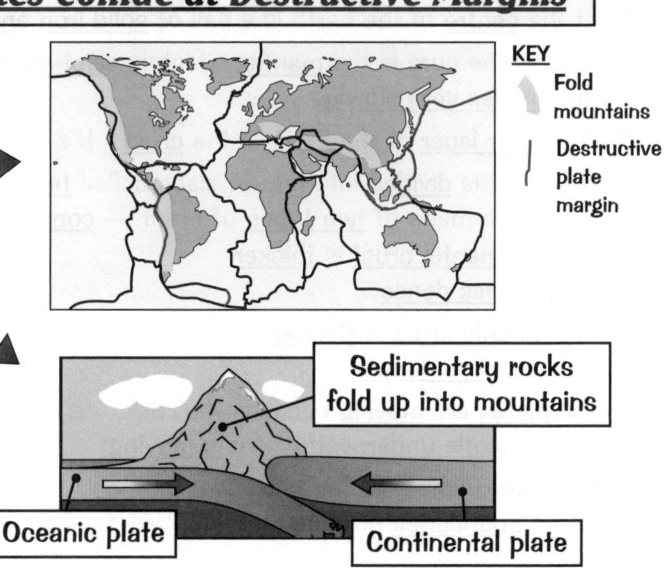

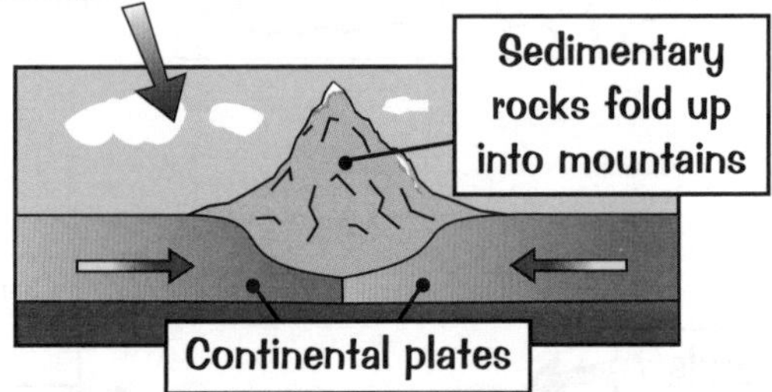

5) Fold mountain areas have lots of very high mountains, which are very rocky with steep slopes. There's often snow and glaciers in the highest bits, and lakes in the valleys between the mountains.

Humans Use Fold Mountain Areas for Lots of Things

FARMING: Higher mountain slopes aren't great for growing crops so they're used to graze animals, e.g. mountain goats. Lower slopes are used to grow crops. Steep slopes are sometimes terraced to make growing crops easier.

HYDRO-ELECTRIC POWER (HEP): Steep-sided mountains and high lakes (to store water) make fold mountains ideal for generating hydro-electric power.

MINING: Fold mountains are a major source of metal ores, so there's a lot of mining going on. The steep slopes make access to the mines difficult, so zig-zag roads have been carved out on the sides of some mountains to get to them.

FORESTRY: Fold mountain ranges are a good environment to grow some types of tree (e.g. conifers). They're grown on the steep valley slopes and are used for things like fuel, building materials, and to make things like paper and furniture.

TOURISM: Fold mountains have spectacular scenery, which attracts tourists. In winter, people visit to do sports like skiing, snowboarding and ice climbing. In summer, walkers come to enjoy the scenery. Tunnels have been drilled through some fold mountains to make straight, fast roads. This improves communications for tourists and people who live in the area as it's quicker to get to places.

Fold-up mountains — you can take them anywhere...

Yep, fold mountains are pretty much what they say they are — mountains made by folding. Humans use them for lots of things — including using them as a handy place to go to the loo. Maybe don't use that one in the exam though.

Fold Mountains — Case Study

Brace yourself for the first case study of many. It's quite a groovy one about the Alps though. Make sure you learn the specific facts for good marks in the exam.

The Alps is a Fold Mountain Range

Location: Central Europe — it stretches across Austria, France, Germany, Italy, Liechtenstein, Slovenia and Switzerland.

Formation: The Alps were formed about 30 million years ago by the collision between the African and European plates.

Tallest peak: Mont Blanc at 4810 m on the Italian-French border.

Population: Around 12 million people.

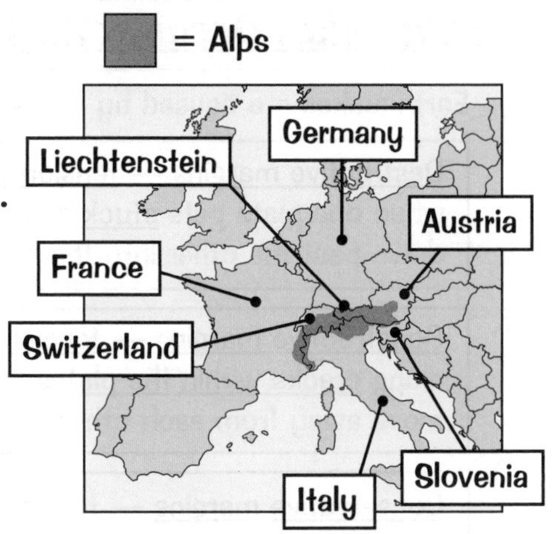

People Use the Alps for Lots of Things

FARMING

1) The steep upland areas are used to farm goats, which provide milk, cheese and meat.
2) Some sunnier slopes have been terraced to plant vineyards (e.g. Lavaux, Switzerland).

TOURISM

1) 100 million tourists visit the Alps each year making tourism a huge part of the economy.
2) 70% of tourists visit the steep, snow covered mountains in the winter for skiing, snowboarding and ice climbing. In the summer tourists visit for walking, mountain biking, paragliding and climbing.
3) New villages have been built to cater for the quantity of tourists, e.g. Tignes in France.
4) Ski runs, ski lifts, cable cars, holiday chalets and restaurants pepper the landscape.

HYDRO-ELECTRIC POWER (HEP)

1) The narrow valleys are dammed to generate HEP, e.g. in the Berne area in Switzerland. Switzerland gets 60% of its electricity from HEP stations in the Alps.
2) The electricity produced is used locally to power homes and businesses. It's also exported to towns and cities further away.

MINING

Salt, iron ore, gold, silver and copper were mined in the Alps, but the mining has declined dramatically due to cheaper foreign sources.

FORESTRY

Scots Pine is planted all over the Alps because it's more resilient to the munching goats, which kill native tree saplings. The trees are logged and sold to make things like furniture.

People Have Adapted to the Conditions in the Alps

1) STEEP RELIEF: Goats are farmed there because they're well adapted to live on steep mountains. Trees and man-made defences are used to protect against avalanches and rock slides.
2) POOR SOILS: Animals are grazed in most high areas as the soil isn't great for growing crops.
3) LIMITED COMMUNICATIONS: Roads have been built over passes (lower points between mountains), e.g. the Brenner Pass between Austria and Italy. It takes a long time to drive over passes and they can be blocked by snow, so tunnels have been cut through the mountains to provide fast transport links. For example, the Lötschberg Base Tunnel has been cut through the Bernese Alps in Switzerland.

Steep relief — the feeling you'll get after you've done your exams...

As with all case studies learn the facts and figures — the examiners go all giddy and throw marks at you when they read them. In the exam you could be asked how people use the Alps and how they've adapted to the conditions there.

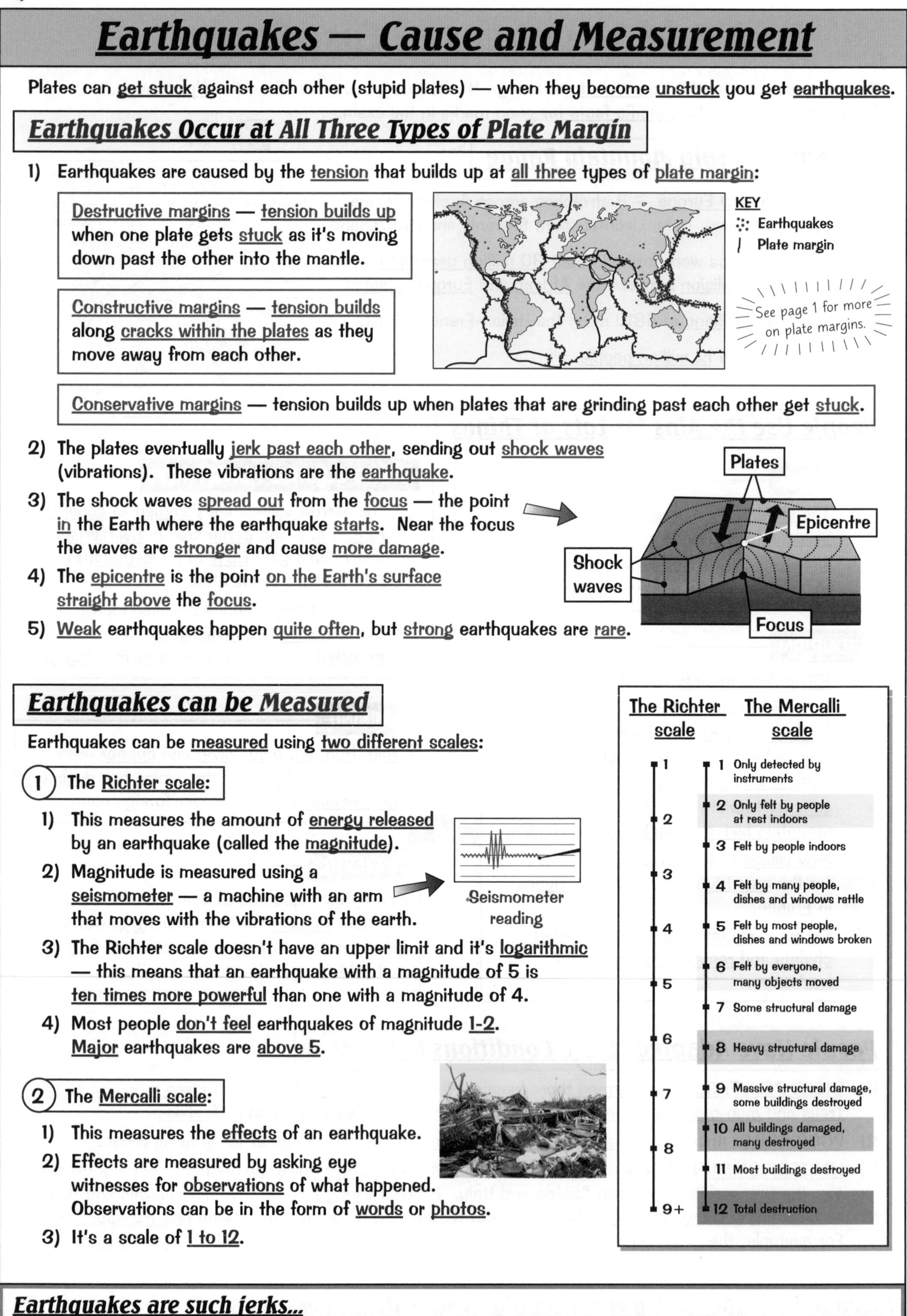

Earthquakes — Cause and Measurement

Plates can get stuck against each other (stupid plates) — when they become unstuck you get earthquakes.

Earthquakes Occur at All Three Types of Plate Margin

1) Earthquakes are caused by the tension that builds up at all three types of plate margin:

Destructive margins — tension builds up when one plate gets stuck as it's moving down past the other into the mantle.

Constructive margins — tension builds along cracks within the plates as they move away from each other.

Conservative margins — tension builds up when plates that are grinding past each other get stuck.

See page 1 for more on plate margins.

2) The plates eventually jerk past each other, sending out shock waves (vibrations). These vibrations are the earthquake.
3) The shock waves spread out from the focus — the point in the Earth where the earthquake starts. Near the focus the waves are stronger and cause more damage.
4) The epicentre is the point on the Earth's surface straight above the focus.
5) Weak earthquakes happen quite often, but strong earthquakes are rare.

Earthquakes can be Measured

Earthquakes can be measured using two different scales:

1 The Richter scale:

1) This measures the amount of energy released by an earthquake (called the magnitude).
2) Magnitude is measured using a seismometer — a machine with an arm that moves with the vibrations of the earth.
3) The Richter scale doesn't have an upper limit and it's logarithmic — this means that an earthquake with a magnitude of 5 is ten times more powerful than one with a magnitude of 4.
4) Most people don't feel earthquakes of magnitude 1-2. Major earthquakes are above 5.

2 The Mercalli scale:

1) This measures the effects of an earthquake.
2) Effects are measured by asking eye witnesses for observations of what happened. Observations can be in the form of words or photos.
3) It's a scale of 1 to 12.

Earthquakes are such jerks...

No one really likes earthquakes — they're pretty tense and they could crack at any moment. They happen at all three types of plate boundary but the big ones don't come round very often — I'd imagine they don't get many invitations.

Impacts of Earthquakes

Earthquakes make jelly wobble — but they have loads of more serious impacts...

Earthquakes have Primary and Secondary Impacts

The primary impacts of an earthquake are the immediate effects of the ground shaking. The secondary impacts are the effects that happen later on. Here are a few examples of the possible impacts:

Primary impacts

1) Buildings and bridges collapse.
2) People are injured or killed by buildings and bridges collapsing.
3) Roads, railways, ports and airports are damaged.
4) Electricity cables are damaged, cutting off supplies.
5) Gas pipes are broken, causing leaks and cutting off supplies.
6) Telephone poles and cables are destroyed.
7) Underground water and sewage pipes are broken, causing leaks and cutting off supplies.

Secondary impacts

Tsunamis are a series of enormous waves caused when huge amounts of water get displaced — there's more about them on page 13.

1) Earthquakes can trigger landslides and tsunamis — these destroy more buildings and cause more injuries and deaths.
2) Leaking gas can be ignited, starting fires.
3) People are left homeless.
4) People may suffer psychological problems if they knew people who died or if they lose their home etc.
5) There's a shortage of clean water and a lack of proper sanitation — this makes it easier for diseases to spread.
6) Roads are blocked or destroyed so aid and emergency vehicles can't get through.
7) Businesses are damaged or destroyed, causing unemployment.

The more settlements built and businesses set up in an area, the greater the impact because there are more people and properties to be affected by an earthquake.

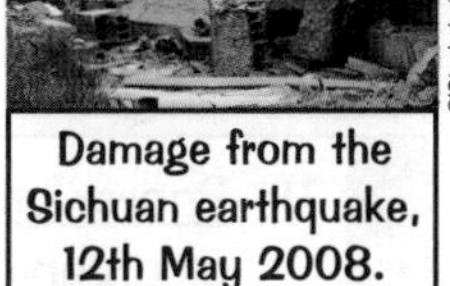

Damage from the Sichuan earthquake, 12th May 2008.

The Impacts are More Severe in Poorer Countries

Here are a few reasons why:

1) There's more low quality housing in poorer countries. Low quality houses are less stable, so they're destroyed more easily by earthquakes.
2) The infrastructure is often worse in poorer countries. Poor quality roads make it harder for emergency services to reach injured people, which leads to more deaths.
3) Poorer countries don't have much money to protect against earthquakes, e.g. by making buildings earthquake proof. They also don't have enough money or resources (e.g. food and emergency vehicles) to react straight away to earthquakes, so more people are affected by secondary impacts.
4) Healthcare is often worse in poorer countries. Many hospitals in poorer countries don't have enough supplies to deal with lots of casualties after an earthquake, so more people die from treatable injuries.

People Continue to Live in the Areas where Earthquakes Happen

Here are some of the reasons why people don't move away from earthquake prone areas, e.g. California:

1) They've always lived there — moving away would mean leaving friends and family.
2) They're employed in the area. If people move they would have to find new jobs.
3) They're confident of support from their government after an earthquake, e.g. to help rebuild houses.
4) Some people think that severe earthquakes won't happen again in the area, so it's safe to live there.

Some people just like living on the edge (of a plate)...

Earthquakes seem pretty exciting to the outsider, but they can be life-threatening for the people that experience them. All the different impacts affect both richer countries and poorer countries, but they're more severe in poorer countries.

Reducing the Impacts of Earthquakes

Unfortunately earthquakes don't have an 'off' button, but the impacts they have can be reduced.

There are Many Ways of Reducing the Impacts of Earthquakes

Prediction

1) It's currently impossible to predict when an earthquake will happen. If you could, it would give people time to evacuate — this would reduce the number of injuries and deaths.
2) There can be clues that an earthquake is about to happen though. For example, lots of small tremors, cracks appearing in rocks and strange animal behaviour (e.g. rats abandoning nests).
3) It's possible to predict where future earthquakes may happen using data from past earthquakes, e.g. mapping where earthquakes have happened shows which places are likely to be affected again — these places can prepare themselves for the impacts of an earthquake.

Building techniques

1) Buildings can be designed to withstand earthquakes, e.g. by using materials like reinforced concrete or building special foundations that absorb an earthquake's energy.
2) Constructing earthquake-proof buildings reduces the number of buildings destroyed by an earthquake, so fewer people will be killed, injured, made homeless and made unemployed.

Planning

1) Future developments, e.g. new shopping centres, can be planned to avoid the areas most at risk from earthquakes. This reduces the number of buildings destroyed by an earthquake.
2) Firebreaks can be made to reduce the spread of fires (a secondary impact, see previous page).
3) Emergency services can train and prepare for disasters, e.g. by practising rescuing people from collapsed buildings and by stockpiling medicine. This reduces the number of people killed.
4) Governments can plan evacuation routes to get people out of dangerous areas quickly and safely after an earthquake. This reduces the number of people killed or injured by things like fires.

Education

1) Governments and other organisations can educate people about what to do if there's an earthquake (e.g. stand in a doorway) and how to evacuate. This reduces deaths.
2) People can be told how to make a survival kit containing things like food, water, a torch, a radio and batteries. The kits reduce the chances of people dying if they're stuck in the area.

Aid

1) Poorer countries that have been affected by earthquakes can receive aid from governments or organisations — it can be things like food, water, money or people (e.g. doctors or rescuers).
2) Aid helps to reduce the impacts, e.g. money aid is used to rebuild homes, reducing homelessness.

Some Strategies are More Sustainable than Others

Sustainable strategies meet the needs of people today without stopping people in the future meeting their needs. A strategy is not sustainable if it's not effective, as it doesn't meet the needs of people today. A strategy is also not sustainable if it's expensive or it harms the environment, as it stops people in the future meeting their needs. Here's a bit on the sustainability of strategies to reduce the impact of earthquakes:

1) Predicting earthquakes is not an effective strategy, so it's not sustainable.
2) The other strategies above are sustainable — they're all effective and environmentally friendly, but some are more sustainable than others.
3) The ones that are more sustainable are basically the more cost-effective ones, e.g. good planning is usually more effective than aid at reducing impacts and it's much cheaper.
4) Some strategies are expensive (e.g. constructing earthquake-proof buildings), but they can be more sustainable than other strategies because in the long term less money and resources are used rebuilding.

I'd reduce the impacts by making everything out of rubber — everything...

If you could predict exactly when an earthquake will happen you'd make yourself a lot of friends, but top scientists think it's just too difficult to do. Anyway, it's worth knowing about the other strategies too and their sustainability.

Earthquakes — Case Studies

And you thought I'd forgotten all about the case studies. Shame on you.

Rich and Poor Parts of the World are Affected Differently

The effects of earthquakes and the responses to them are different in different parts of the world. A lot depends on how wealthy the part of the world is. I'd bet my budgie they'll want you to compare two earthquakes in a rich and a poor part of the world in the exam.

Earthquake in a rich part of the world:

Place: L'Aquila, Italy
Date: 6th April, 2009
Size: 6.3 on the Richter Scale
Cause: Movement along a crack in the plate at a destructive margin.
Cost of damage: Around $15 billion

Earthquake in a poor part of the world:

Place: Kashmir, Pakistan
Date: 8th October, 2005
Size: 7.6 on the Richter Scale
Cause: Movement along a crack in the plate at a destructive margin.
Cost of damage: Around $5 billion

	Rich (L'Aquila, Italy)	Poor (Kashmir, Pakistan)
Preparation	• There are laws on construction standards, but some modern buildings hadn't been built to withstand earthquakes. • Italy has a Civil Protection Department that trains volunteers to help with thing like rescue operations.	• No local disaster planning was in place. • Buildings were not designed to be earthquake resistant. • Communications were poor. There were few roads and they were badly constructed.
Primary effects	• Around 290 deaths, mostly from collapsed buildings. • Hundreds of people were injured. • Thousands of buildings were damaged or destroyed. • A bridge near the town of Fossa collapsed, and a water pipe was broken near the town of Paganica.	• Around 80 000 deaths, mostly from collapsed buildings. • Hundreds of thousands of people injured. • Entire villages and thousands of buildings were destroyed. • Water pipelines and electricity lines were broken, cutting off supply.
Secondary effects	• Aftershocks hampered rescue efforts and caused more damage. • Thousands of people were made homeless. • Fires in some collapsed buildings caused more damage. • The broken water pipe near the town of Paganica caused a landslide.	• Landslides buried buildings and people. They also blocked access roads and cut off water supplies, electricity supplies and telephone lines. • Around 3 million people were made homeless. • Diarrhoea and other diseases spread due to little clean water. • Freezing winter conditions shortly after the earthquake caused more casualties and meant rescue and rebuilding operations were difficult.
Immediate response	• Camps were set up for homeless people with water, food and medical care. • Ambulances, fire engines and the army were sent in to rescue survivors. • Cranes and diggers were used to remove rubble. • International teams with rescue dogs were sent in to look for survivors. • Money was provided by the government to pay rent, and gas and electricity bills were suspended.	• Help didn't reach many areas for days or weeks. People had to be rescued by hand without any equipment or help from emergency services. • Tents, blankets and medical supplies were distributed within a month, but not to all areas affected. • International aid and equipment such as helicopters and rescue dogs were brought in, as well as teams of people from other countries.
Long-term response	• The Italian Prime Minister promised to build a new town to replace L'Aquila as the capital of the area. • An investigation is going on to look into why the modern buildings weren't built to withstand earthquakes.	• Around 40 000 people have been relocated to a new town from the destroyed town of Balakot. • Government money has been given to people whose homes had been destroyed so they can rebuild them themselves. • Training has been provided to help rebuild more buildings as earthquake resistant. • New health centres have been set up in the area.

Run away, run away — my immediate response to earthquakes...

The amount of damage an earthquake does, and the number of people that get hurt, is different in different parts of the world. In the exam you might be asked to compare an earthquake from a rich part of the world with one from a poor part.

Volcanoes

Volcanoes usually look like mountains... until they explode and throw molten rock everywhere. Honestly, they've got such a temper, though I'd probably be in a bad mood if I was sitting on a load of magma.

Volcanoes are Found at Destructive and Constructive Plate Margins

1) At destructive plate margins the oceanic plate goes under the continental plate because it's more dense. (This also creates an ocean trench.):
   - The oceanic plate moves down into the mantle, where it's melted and destroyed.
   - A pool of magma forms.
   - The magma rises through cracks in the crust called vents.
   - The magma erupts onto the surface (where it's called lava) forming a volcano.
2) At constructive margins the magma rises up into the gap created by the plates moving apart, forming a volcano.
3) Some volcanoes also form over parts of the mantle that are really hot (called hotspots), e.g. in Hawaii.

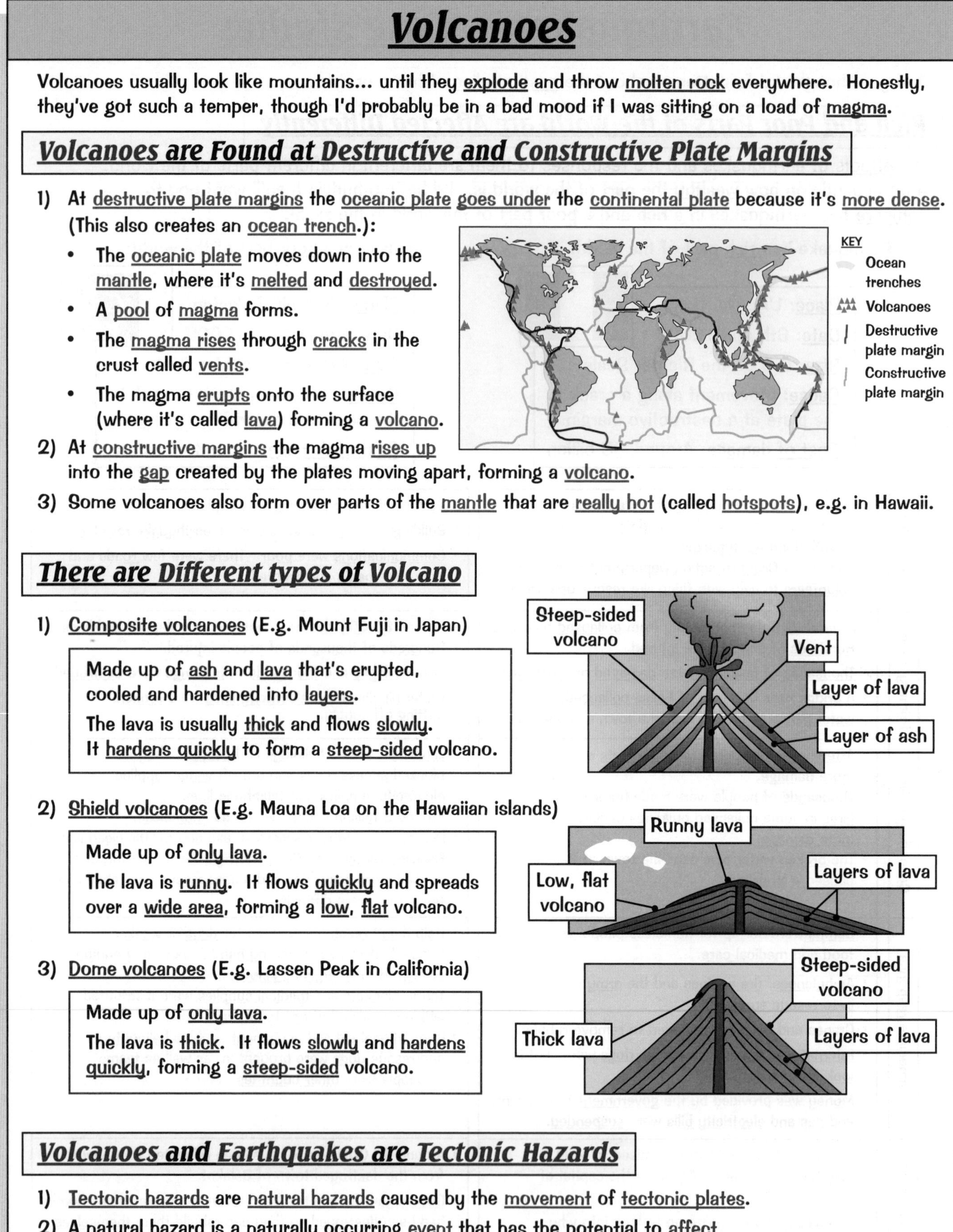

There are Different types of Volcano

1) Composite volcanoes (E.g. Mount Fuji in Japan)

   Made up of ash and lava that's erupted, cooled and hardened into layers.
   The lava is usually thick and flows slowly.
   It hardens quickly to form a steep-sided volcano.

2) Shield volcanoes (E.g. Mauna Loa on the Hawaiian islands)

   Made up of only lava.
   The lava is runny. It flows quickly and spreads over a wide area, forming a low, flat volcano.

3) Dome volcanoes (E.g. Lassen Peak in California)

   Made up of only lava.
   The lava is thick. It flows slowly and hardens quickly, forming a steep-sided volcano.

Volcanoes and Earthquakes are Tectonic Hazards

1) Tectonic hazards are natural hazards caused by the movement of tectonic plates.
2) A natural hazard is a naturally occurring event that has the potential to affect people's lives or property.
3) When natural hazards do affect people's lives or property they're called natural disasters.

Studying volcanoes — what a blast...

There are different types of volcano, but they've all got one thing in common — throwing up lava without bothering to rush to the toilet. You'll never have to draw a map like the one above, but you should have an idea of where volcanoes occur.

Impacts of Volcanoes

Living by a volcano means you have to put up with quite a lot — lava, exploding mountains, flying monkeys*. Not really my cup of tea, but plenty of people keep living near them. *Oops, no flying monkeys, I just got a bit carried away.

Lots of People Live Close to Volcanoes

The reasons why people continue to live around volcanoes despite the hazards are exactly the same as why people keep living in areas prone to earthquakes (see page 5). But there are a few reasons why people choose to live close to volcanoes:

1) The soil around volcanoes is fertile because it's full of minerals from volcanic ash and lava. This makes it good for growing crops, which attracts farmers.
2) Volcanoes are tourist attractions — loads of tourists visit volcanoes so lots of people live around volcanoes to work in the tourist industry.
3) Volcanoes are a source of geothermal energy, which can be used to generate electricity. So people live around volcanoes to work at power stations.

Mount St. Helens (USA), Nevado del Ruiz (Colombia) and Mount Etna (Sicily) are volcanoes that have erupted recently.

Volcanic Eruptions have Primary and Secondary Impacts

The primary impacts of a volcanic eruption are the immediate effects of a volcano spewing out lava, ash, rocks and gas (e.g. carbon dioxide and sulphur dioxide), as well as pyroclastic flows. The secondary impacts are the effects that happen later on.

Pyroclastic flows are extremely fast moving flows of ash, rock and gas that move down the sides of a volcano.

Here are a few examples of the possible impacts:

Primary impacts

1) Buildings and roads are destroyed by lava flows and pyroclastic flows — buildings also collapse if enough ash falls on them.
2) People and animals are injured or killed, mainly by pyroclastic flows but also by lava flows and falling rocks.
3) Crops are damaged and water supplies are contaminated when ash falls on them.
4) People, animals and plants are suffocated by carbon dioxide.

BERNHARD EDMAIER / SCIENCE PHOTO LIBRARY

This was the city of Plymouth in Montserrat (a poorer country) — it was buried under ash and mud after a volcanic eruption in 1997.

Secondary impacts

1) Mudflows (also called lahars) form when volcanic material mixes with water, e.g. from heavy rainfall or snow melt. Mudflows cause loads more destruction, deaths and injuries.
2) Fires are started by lava flows and pyroclastic flows, which then spread.
3) People may suffer psychological problems if they knew people who died or if they lose their home etc.
4) People are left homeless.
5) There's a shortage of food because crops are damaged.
6) There's a shortage of clean water.
7) Roads are blocked or destroyed so aid and emergency vehicles can't get through.
8) Businesses are damaged or destroyed, causing unemployment.
9) Sulphur dioxide released into the atmosphere causes acid rain.

The impacts of volcanic eruptions are more severe in poorer countries than in richer countries for exactly the same reasons why earthquakes are more severe in poorer countries — have a look back at page 5.

Secondary impact — when a meteor hits your school...

Now that's a great idea for a disaster movie — I wonder if I can get Kevin Spacey... Oops, sorry, I got carried away again. Volcanic eruptions cause various nasty impacts and you should be clear on which are primary and which are secondary.

Reducing the Impacts of Volcanoes

It's impossible to stop a volcano erupting — not even with a humongous bath plug. But a few people have come up with ideas to reduce the impacts of volcanic eruptions as much as possible, which is nice of them.

There are Many Ways of Reducing the Impacts of Volcanic Eruptions

Prediction

1) Unlike earthquakes, it's possible to roughly predict when a volcanic eruption will happen. Scientists can monitor the tell-tale signs that come before a volcanic eruption.
2) Things such as tiny earthquakes, escaping gas, and changes in the shape of the volcano (e.g. bulges in the land where magma has built up under it) all mean an eruption is likely.
3) Predicting when a volcano is going to erupt gives people time to evacuate — this reduces the number of injuries and deaths.

Uh oh...

Planning

1) Future developments, e.g. new houses, can be planned to avoid the areas most at risk from volcanic eruptions. This reduces the number of buildings destroyed by an eruption.
2) Firebreaks can be made to reduce the spread of fires.
3) Emergency services can train and prepare for disasters, e.g. by practising setting up emergency camps for homeless people. This reduces the number of people killed.
4) Governments can plan evacuation routes to get people away from the volcano quickly and safely. This reduces the number of people injured or killed by things like pyroclastic flows or mudflows.

Building techniques

1) Buildings can't be designed to withstand lava flows or pyroclastic flows, but they can be strengthened so they're less likely to collapse under the weight of falling ash.
2) The lava from some volcanoes can be diverted away from buildings using barriers.
3) Both of these reduce the number of buildings destroyed, so fewer people will be killed, injured, made homeless and made unemployed.

Education

1) Governments and other organisations can educate people about how to evacuate if a volcano erupts. This helps people get out of danger quickly and safely, which reduces deaths.
2) People can be told how to make a survival kit containing things like food, water, a torch, a radio, batteries and dust masks. The kits reduce the chance of people dying if they're stuck in the area.

Aid

1) Poorer countries that have been affected by a volcanic eruption can receive aid from governments or organisations — it can be things like food, water, money or people (e.g. doctors).
2) Aid helps to reduce the impacts, e.g. food aid stops people going hungry.

Some Strategies are More Sustainable than Others

There's a definition of a sustainable strategy on page 6. Have a read of this bit about the sustainability of strategies to reduce the impact of volcanic eruptions:

1) All of the strategies are sustainable because they're all effective and environmentally friendly.
2) Some are more cost-effective than others though, so are more sustainable.
3) Predicting eruptions needs special equipment and trained scientists, which makes it expensive, but if it's accurate it saves a lot of lives.
4) Building techniques can be very expensive, but can save money if they stop building destruction.

It's impossible to stop a volcano erupting — have we tried asking nicely...

Please don't try reasoning with volcanoes by the way — they won't listen. I'm afraid it'll just have to be revision for you. While we're on the subject of revision, sustainability pops up a lot, so it's worth making sure you understand what it means.

Volcanic Eruption — Case Study

It's that time again, yep, case study time. Cram your memory full of these facts (the cause of the eruption, impacts and responses — don't think just learning the name and date will cut it).

The Soufrière Hills Volcano in Montserrat Erupted in 1997

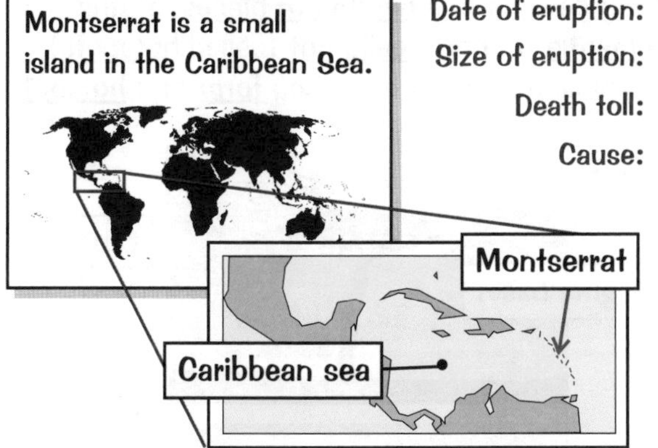

Date of eruption: June 25th 1997 (small eruptions started in July 1995).

Size of eruption: Large — 4-5 million m^3 of rocks and gas released.

Death toll: 19 killed

Cause:
1) Montserrat is above a destructive plate margin, where the North American plate is being forced under the Caribbean plate.
2) Magma rose up through weak points under the Soufrière hills forming an underground pool of magma.
3) The rock above the pool collapsed, opening a vent and causing the eruption.

There were Primary and Secondary Impacts

Primary impacts

1) Large areas were covered with volcanic material — the capital city Plymouth was buried under 12 m of mud and ash.
2) Over 20 villages and two thirds of homes on the island were destroyed by pyroclastic flows (fast-moving clouds of super-heated gas and ash).
3) Schools, hospitals, the airport and the port were destroyed.
4) Vegetation and farmland were destroyed.
5) 19 people died and 7 were injured.

Secondary impacts

1) Fires destroyed many buildings including local government offices, the police headquarters and the town's central petrol station.
2) Tourists stayed away and businesses were destroyed, disrupting the economy.
3) Population decline — 8000 of the island's 12 000 inhabitants have left since the eruptions began in 1995.
4) Volcanic ash from the eruption has improved soil fertility.
5) Tourism on the island is now increasing as people come to see the volcano.

Immediate responses

1) People were evacuated from the south to safe areas in the north.
2) Shelters were built to house evacuees.
3) Temporary infrastructure was also built, e.g. roads and electricity supplies.
4) The UK provided £17 million of emergency aid (Montserrat's an overseas territory of the UK).
5) Local emergency services provided support units to search for and rescue survivors.

Long-term responses

1) A risk map was created and an exclusion zone is in place. The south of the island is off-limits while the volcano is still active.
2) The UK has provided £41 million to develop the north of the island — new docks, an airport and houses have been built in the north.
3) The Montserrat Volcano Observatory has been set up to try and predict future eruptions.

Learn the Eruption and without Interruption learn the Eruption Disruption...

The eruption on Montserrat wasn't all bad, well... it was mostly bad, but there were some positive impacts — the ash improved the soil fertility and lots of tourists now go to gawp at the volcano. Every cloud has a silver lining and all that.

Supervolcanoes

Normal volcanoes just not disastrous enough for you? What you need is a supervolcano. The same lava and ash combo, but global destruction thrown in. It's a brand new topic so an exam question's likely.

Supervolcanoes are Massive Volcanoes

Supervolcanoes are much bigger than standard volcanoes. They develop in a handful of places around the globe — at destructive plate margins or over parts of the mantle that are really hot (called hotspots), e.g. Yellowstone National Park in the USA is on top of a supervolcano. Here's how they form at a hotspot:

1) Magma rises up through cracks in the crust to form a large magma basin below the surface. The pressure of the magma causes a circular bulge on the surface several kilometres wide.
2) The bulge eventually cracks, creating vents for lava to escape through. The lava erupts out of the vents causing earthquakes and sending up gigantic plumes of ash and rock.
3) As the magma basin empties, the bulge is no longer supported so it collapses — spewing up more lava.
4) When the eruption's finished there's a big crater (called a caldera) left where the bulge collapsed. Sometimes these get filled with water to form a large lake, e.g. Lake Toba in Indonesia.

Before: Bulge, Crust, Magma basin, Mantle

During: Vents

After: Caldera, Collapsed basin

You need to know the characteristics of a supervolcano:

- Flat (unlike normal volcanoes, which are mountains).
- Cover a large area (much bigger than normal volcanoes).
- Have a caldera (normal volcanoes may or may not have a caldera — they might just have a crater).

When a Supervolcano Erupts there will be Global Consequences

Fortunately there are only a few supervolcanoes and an eruption hasn't happened for tens of thousands of years, e.g. the last one to erupt was the Lake Toba supervolcano 74 000 years ago.
When there is an eruption though, it's predicted that an enormous area will be affected:

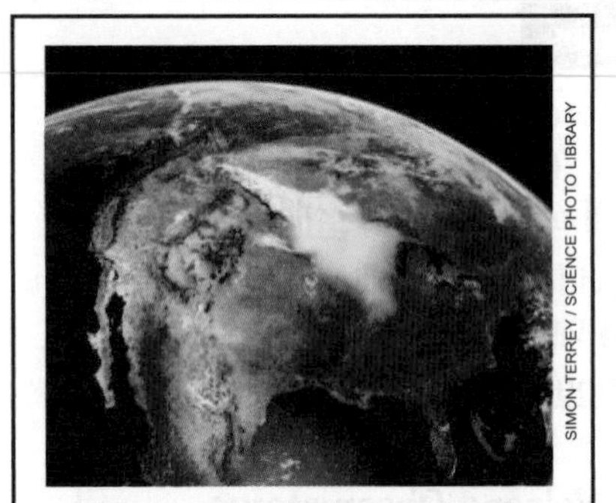

The predicted plume of ash from a supervolcanic eruption in Yellowstone National Park.

1) A supervolcanic eruption will throw out thousands of cubic kilometres of rock, ash and lava (much more than normal volcanoes, which usually produce a couple of cubic kilometres).
2) A thick cloud of super-heated gas and ash will flow at high speed from the volcano, killing, burning and burying everything it touches. Everything within tens of miles will be destroyed.
3) Ash will shoot kilometres into the air and block out almost all daylight over whole continents. This can trigger mini ice ages as less heat energy from the sun gets to Earth.
4) The ash will also settle over hundreds of square kilometres, burying fields and buildings (ash from normal volcanoes usually covers a couple of square kilometres).

The apocalypse is nigh — run away from the flat areas that look like a caldera...

Supervolcanoes are very different beasts from dome or shield volcanoes. They form in a slightly different way and have way bigger impacts. Oh, and you may want to reconsider your decision to move to Yellowstone National Park.

Tsunamis — Case Study

As if volcanoes and earthquakes weren't bad enough, if they happen out at sea they can cause tsunamis. A tsunami is a series of enormous waves caused when huge amounts of water get displaced.

An Earthquake caused a Tsunami in the Indian Ocean in 2004

1) There's a destructive plate margin along the west coast of Indonesia in the Indian Ocean.
2) On 26th December 2004 there was an earthquake off the west coast of the island of Sumatra measuring around 9.1 on the Richter scale.
3) The plate that's moving down into the mantle cracked and moved very quickly, which caused a lot of water to be displaced. This triggered a tsunami with waves up to 30 m high.

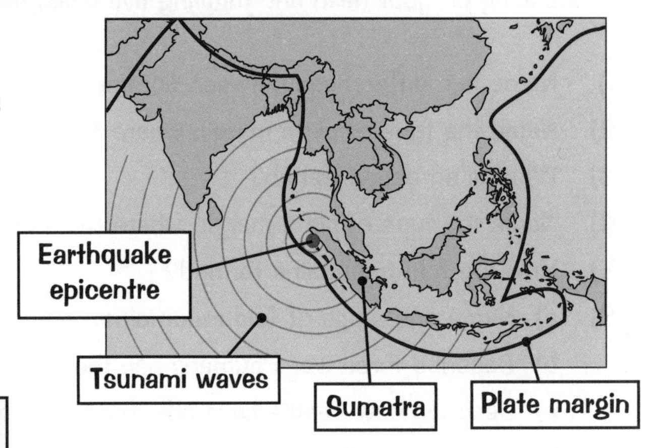

The Tsunami Affected Many Countries

The Indian Ocean tsunami was one of the most destructive natural disasters that's ever happened. It affected most countries bordering the Indian Ocean, e.g. Indonesia, Thailand, India and Sri Lanka. The effects of the tsunami were so bad because there was no early warning system:

1) Around 230 000 people were killed or are still missing.
2) Whole towns and villages were destroyed — over 1.7 million people lost their homes.
3) The infrastructure (things like the roads, water pipes and electricity lines) of many countries was severely damaged.
4) 5-6 million people needed emergency food, water and medical supplies.
5) There was massive economic damage. Millions of fishermen lost their livelihoods, and the tourism industry suffered because of the destruction and because people were afraid to go on holiday there.
6) There was massive environmental damage. Salt from the seawater has meant plants can't grow in many areas. Mangroves, coral reefs, forests and sand dunes were also destroyed by the waves.

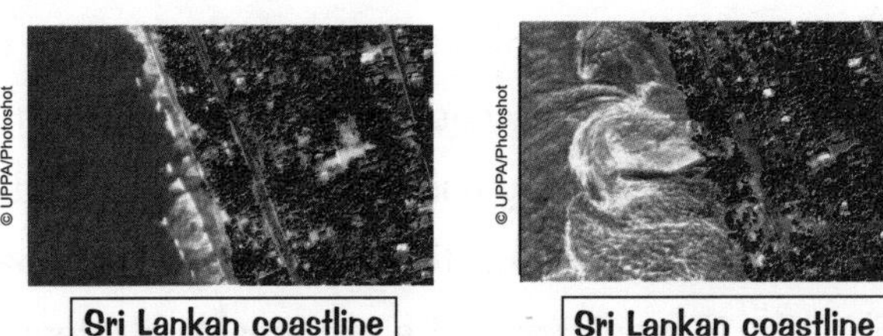

Sri Lankan coastline before the tsunami

Sri Lankan coastline during the tsunami

The Response Involved a lot of International Aid

Short-term responses:

1) Within days hundreds of millions of pounds had been pledged by foreign governments, charities, individuals and businesses to give survivors access to food, water, shelter and medical attention.
2) Foreign countries sent ships, planes, soldiers and teams of specialists to help rescue people, distribute food and water and begin clearing up.

Long-term responses:

1) Billions of pounds have been pledged to help re-build the infrastructure of the countries affected.
2) As well as money, programmes have been set up to re-build houses and help people get back to work.
3) A tsunami warning system has been put in place in the Indian Ocean.
4) Disaster management plans have been put in place in some countries. Volunteers have been trained so that local people know what to do if a tsunami happens again.

Tsunamis can be utterly devastating...

Crikey, tsunamis can wreak as much havoc as the earthquakes or volcanoes that cause them. Still, that's the end of this danger-filled section, and I think I need a bit of a lie down. However, there's the matter of a few questions...

Revision Summary for Section 1

What a cracking section, earth-shattering one might even say. It may be a section about disasters, but get a load of these questions down you and there won't be any kind of disaster in the exam. I know it looks like there's a lot of stuff here, but you'll be surprised how much you just learnt. Try them out a few at a time, then check the answers on the pages. Once you can answer them all standing on your head and juggling five balls, move on to the next section. It rocks, trust me...

1) Name two differences between continental plates and oceanic plates.
2) Name the type of plate margin where two plates are moving towards each other.
3) What is an ocean trench?
4) Name the type of plate margin where two plates are moving sideways against each other.
5) How are fold mountains formed?
6) a) Name one range of fold mountains.
   b) Describe three ways humans use the area.
   c) Describe how people have adapted to the conditions of the area.
7) What causes earthquakes?
8) What's the point in the Earth called where an earthquake starts?
9) What does the Richter scale measure?
10) What is the difference between the primary impacts and the secondary impacts of an earthquake?
11) Give three examples of a secondary impact of an earthquake.
12) Describe two reasons why the impacts of an earthquake are more severe in poorer countries.
13) Describe one way buildings can be designed to withstand an earthquake.
14) Describe how education can reduce the impacts of earthquakes.
15) Why is prediction not a sustainable strategy for reducing the impacts of an earthquake?
16) a) Name a richer country and a poorer country where an earthquake caused a disaster.
    b) Describe two primary impacts and two secondary impacts of each disaster.
    c) Describe two responses to each earthquake.
17) Name the two types of plate margin that volcanoes are found at.
18) Which type of volcano is made up of layers of ash and lava? Name an example.
19) Which type of volcano is formed when the lava is runny? Name an example.
20) Give two reasons why people choose to live close to volcanoes.
21) Give three examples of a primary impact of a volcanic eruption.
22) How do scientists try to predict volcanic eruptions?
23) Describe two planning strategies that reduce the impact of a volcanic eruption.
24) a) Name a volcanic eruption and state when and where it happened.
    b) Describe two negative primary and two negative secondary impacts of the eruption.
    c) Give two positive impacts of the eruption.
    d) Give two immediate responses and two long-term responses.
25) Where do supervolcanoes form?
26) Give one way that a supervolcanic eruption is different from a volcanic eruption.
27) Give one predicted effect of a supervolcanic eruption.
28) What causes a tsunami?
29) Give an example of a tsunami and describe two of its effects.

Types of Rock

There are three types of rock — hard, heavy and punk... erm, I mean igneous, sedimentary and metamorphic. Rock type depends on how the rock was formed.

Igneous Rocks are Formed from Magma that's Cooled Down

All igneous rocks are formed when molten rock (magma) from the mantle cools down and hardens. There are two types depending on where the magma has cooled down:

The mantle is a layer of semi-molten rock deep in the Earth.

1 INTRUSIVE igneous rocks, e.g. granite

1) These form when magma cools down below the Earth's surface.
2) The magma cools down very slowly, forming large crystals that give the rocks a coarse texture.
3) Large domes of cooled magma form domes of igneous rock called batholiths.
4) Where the magma has flowed into gaps in the surrounding rock it forms dykes (in vertical gaps) and sills (in horizontal gaps).

2 EXTRUSIVE igneous rocks, e.g. basalt

1) These form when magma cools down after it's erupted from a volcano onto the Earth's surface.
2) The magma cools down very quickly, forming small crystals that give the rocks a fine texture.

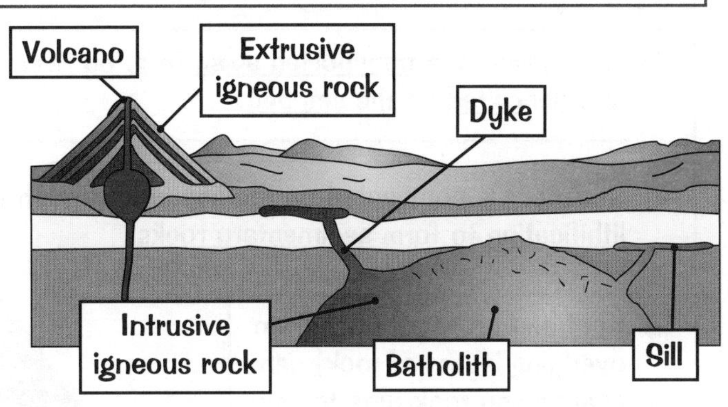

Sedimentary Rocks are Formed from Compacted Sediment

Sedimentary rocks are formed when layers of sediment are compacted together until they become solid rock. The process of compaction is called lithification. Here are a couple of examples:

1) Carboniferous limestone and chalk are formed from calcium carbonate. Layers of tiny shells and skeletons of dead sea creatures are deposited on the sea bed and compacted together over time.
2) Clays and shales are made from mud and clay minerals. The particles have been eroded from older rocks, deposited in layers on lake or sea beds then compacted together.

Sedimentary rocks often contain fossils.

Metamorphic Rocks are Formed by Heat and Pressure

Metamorphic rocks are formed when other rocks (igneous, sedimentary or older metamorphic rocks) are changed by heat and pressure:

1) Rocks deep in the Earth are changed by the pressure from the weight of the material above them.
2) When tectonic plates collide, rocks are changed by the massive heat and pressure that builds up.
3) Magma from the mantle heats the rocks in the crust, causing them to change.

The new rocks are harder and more compact, e.g. limestone becomes marble and clay becomes slate.

This map shows the location of the three rock types in the UK.

*
Igneous rocks
Sedimentary rocks
Metamorphic rocks

Stuck between a rock and a hard exam...

It's easy to get the different rocks mixed up, but don't panic yet. Use the names to help you remember how they're formed — sedimentary is from sediment, metamorphic is rock that's morphed (changed) and igneous is, well, the other one.

The Rock Cycle

You might think once a rock exists then that's it — but weirdly they can change from one type into another.

The Formation of all Rock Types is Linked by the Rock Cycle

The rock cycle shows how igneous, sedimentary and metamorphic rocks are formed, and how one type is changed into another:

1) Weathering (the breakdown of rocks) of all three rock types creates loose sediment.

2) This makes it easier for erosion (the removal of rock) to occur.

3) The sediment is transported away (e.g. by rivers) and deposited on the sea bed.

4) Sediment is compacted on the sea bed through lithification to form sedimentary rocks.

5) Heat and pressure (e.g. from overlying layers of rock) can change any rock type to new metamorphic rock.

6) Melting of any rock type (e.g. in the mantle) creates magma. When magma cools, igneous rocks are formed.

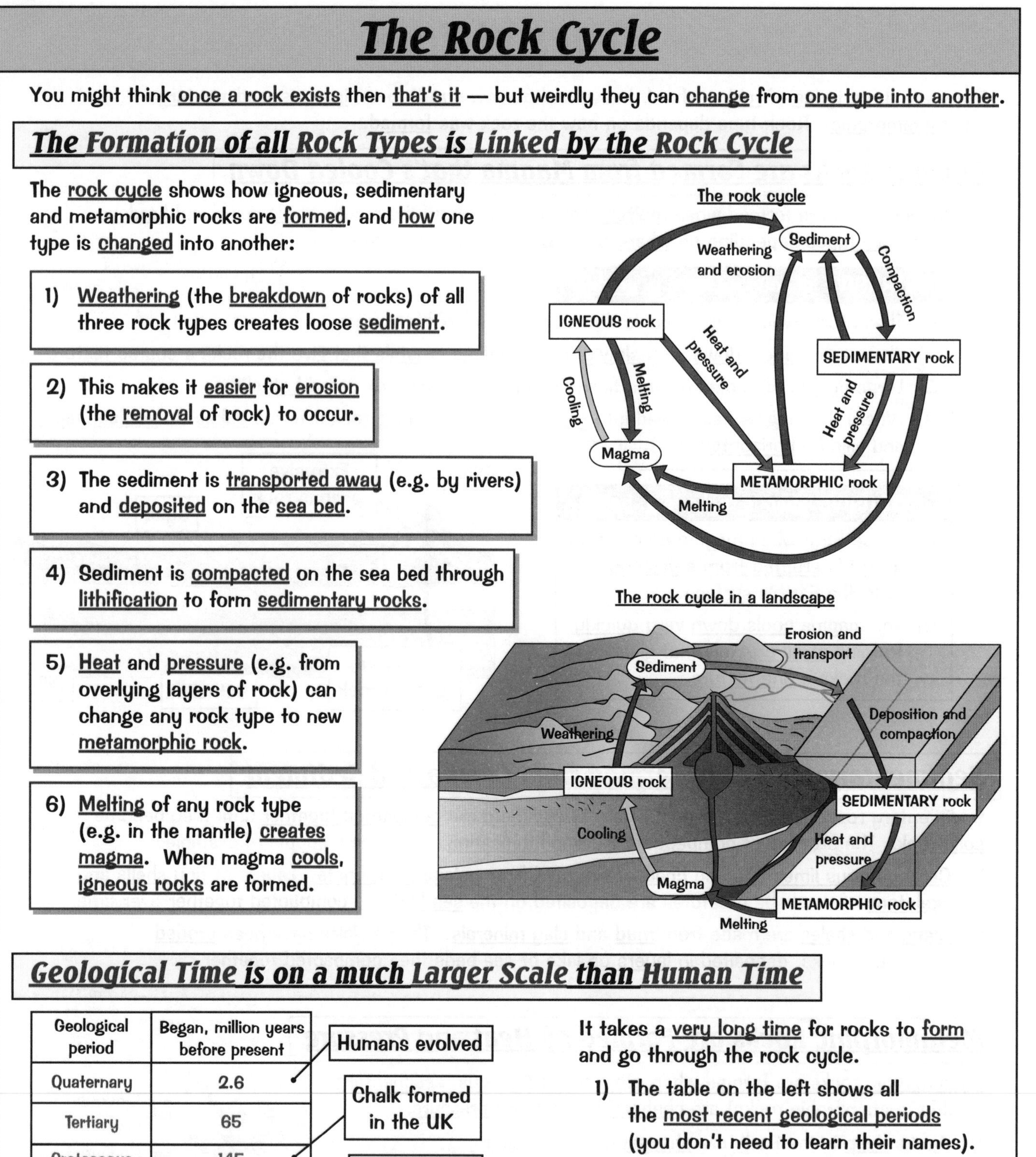

Geological Time is on a much Larger Scale than Human Time

Geological period	Began, million years before present
Quaternary	2.6
Tertiary	65
Cretaceous	145
Jurassic	215
Triassic	245
Permian	285
Carboniferous	360
Devonian	410
Silurian	440
Ordovician	505
Cambrian	585

Humans evolved (Quaternary)

Chalk formed in the UK (Cretaceous)

Clay formed in the UK (Jurassic)

Carboniferous limestone formed in the UK (Carboniferous)

Granite formed in the UK (Ordovician)

These are the major periods when each of these rocks formed — some of them formed at other times too.

It takes a very long time for rocks to form and go through the rock cycle.

1) The table on the left shows all the most recent geological periods (you don't need to learn their names).
2) The rocks found in the UK today were all formed many millions of years ago.
3) Humans (Homo sapiens) have only been around for the last 200 000 years, so the rock cycle is on a totally different time scale to human time.

Rock cycles — really heavy and uncomfortable to sit on...

I feel pretty small and a lot less worried about how old I'll be on my next birthday now — these rocks really are old codgers. Have a go at scribbling down the rock cycle and explaining how one rock type can be changed into another.

Weathering

Weathering is the breakdown of rocks where they are (the material created doesn't get taken away like with erosion). There are three types — mechanical, chemical and biological.

Mechanical Weathering — Rocks are Broken Down by Physical Processes

Mechanical weathering is the breakdown of rocks without changing their chemical composition. Here are two types of mechanical weathering:

FREEZE-THAW weathering

1) In some areas (e.g. upland Britain in winter), the temperature is above 0 °C during the day, and below 0 °C at night.
2) During the day, water gets into cracks in rocks, e.g. granite.
3) At night, the water freezes and expands, which puts pressure on the rock.
4) The water thaws the next day, releasing the pressure, then refreezes the next night.
5) Repeated freezing and thawing widens the cracks and causes the rock to break up.

Day
Night
Water gets into crack
Water freezes and expands
Rock eventually breaks up

EXFOLIATION weathering

1) Some areas have a big daily temperature range, e.g. deserts can be 40 °C in the day and 5 °C at night.
2) Each day the surface layers of rock heat up and expand faster than the inner layers.
3) At night the surface layers cool down and contract faster than the inner layers
4) This creates pressure within the rock and causes thin surface layers to peel off.

Day
Night
Surface layers heat up faster
Surface layers cool down faster
Surface layers peel off

Exfoliation is also known as onion skin weathering.

Chemical Weathering — Rocks are Broken Down by Being Dissolved

Chemical weathering is the breakdown of rocks by changing their chemical composition. Here are two types of chemical weathering:

SOLUTION weathering

1) Some minerals that make up rocks are soluble in water, e.g. rock salt.
2) The minerals dissolve in rainwater, breaking the rock down.

CARBONATION weathering

1) Rainwater has carbon dioxide dissolved in it, which makes it a weak carbonic acid.
2) Carbonic acid reacts with rocks that contain calcium carbonate, e.g. carboniferous limestone, so the rocks are dissolved by the rainwater.

Biological Weathering — Rocks are Broken Down by Plants and Animals

Biological weathering is the breakdown of rocks by living things:

1) Plant roots break down rocks by growing into cracks on their surfaces and pushing them apart.
2) Burrowing animals may loosen small amounts of rock material.

This weathering stuff cracks me up...

So, exfoliating isn't just something that happens at a beauty parlour. Rocks need younger looking, more plumped up skin too. If you know about the three types of weathering, your skin will always be smooth and wrinkle-free. Honest.

Rocks and Landscapes

The type of rock in an area affects the type of landscape that forms.

Granite Landscapes have Tors and Moorland

1) Granite has lots of joints (cracks) which aren't evenly spread (they're closer together in some bits).
2) Freeze-thaw and chemical weathering wear down the parts of the rock with lots of joints faster because there are more cracks for water to get into.
3) Sections of granite that have fewer joints are weathered more slowly than the surrounding rock and stick out at the surface forming tors.

4) Granite is also impermeable — it doesn't let water through.
5) This creates moorlands — large areas of waterlogged and acidic soil, with low-growing vegetation.

Chalk and Clay Landscapes have Escarpments and Vales

1) Horizontal layers of chalk and clay are sometimes tilted diagonally by earth movements.
2) The clay is less resistant than the chalk so is eroded faster.
3) The chalk is left sticking out forming escarpments (hills). Where the clay has been eroded it forms vales — wide areas of flat land.
4) Escarpments have steep slopes (called a scarp slope) on one end, and gentle slopes (dip slope) on the other.
5) Chalk is an aquifer — a permeable rock that stores water.
6) Water flows through the chalk and emerges where the chalk meets impermeable rock (e.g. clay). Where the water emerges is called a spring line.
7) Areas of chalk can also have dry valleys — valleys that don't have a river or stream flowing in them because the water is flowing underground.

Spring line
Dry valley
Escarpment
Aquifer
Clay
Vale
Chalk
Daisy

Escarpments are also called cuestas.

Carboniferous Limestone forms Surface and Underground Features

Rainwater slowly eats away at carboniferous limestone through carbonation weathering (see page 17). Most weathering happens along joints in the rock, creating some spectacular features:

1) Limestone pavements are flat areas of limestone with blocks separated by weathered-down joints.
2) Swallow holes are weathered holes in the surface.
3) Caverns form beneath swallow holes where the limestone has been deeply weathered.
4) Limestone gorges are steep sided gorges formed when caverns collapse.

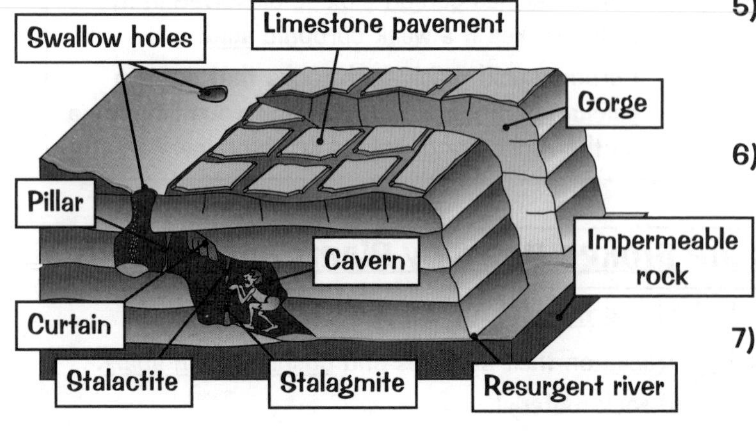

5) Limestone is permeable, so limestone areas also have dry valleys and resurgent rivers (rivers that pop out at the surface when limestone is on top of impermeable rock).
6) Water seeping through limestone contains dissolved minerals. When the water drips into a cavern the minerals solidify and build up over time to produce stalactites (on the ceiling) and stalagmites (on the ground).
7) When stalagmites and stalactites meet in the middle they form a pillar. When water flows in as a sheet a curtain builds up.

Shut your swallow hole and get on with it...

There's a fair bit of information on this page, I'll grant you that, but it's all pretty straightforward stuff. Scribble down the diagrams and label them, then see if you can explain how each landform you've labelled is made.

Using Landscapes — Case Studies

Find out here how people use rock landscapes for resources, farming, tourism and water supplies.

Granite Areas are Used for Stone, Tourism, Farming and Water

1) Granite is quarried and used as a building stone for things like flooring and worktops.
2) Tourists are attracted to the features of granite landscapes, e.g. tors and moorland (see previous page).
3) There are opportunities for rearing livestock on granite areas, but the land isn't great for arable farming (growing crops) or dairy farming because the soils are acidic and waterlogged.
4) Granite's impermeable so granite areas are good places to build reservoirs.

Case Study — Dartmoor, Devon

1) Dartmoor has lots of granite quarries, e.g. Meldon Quarry. Granite from Dartmoor was used to build Nelson's Column.
2) Millions of people visit Dartmoor every year to enjoy the features of the granite landscape, e.g. Bowerman's Nose and Hound Tor are popular attractions.
3) In 2000, there were over 290 000 hectares of land used for rearing livestock (only 900 hectares of land were used for arable farming).
4) There are 8 reservoirs in Dartmoor, e.g. Burrator Reservoir supplies water to Plymouth.

Bowerman's Nose

Chalk and Clay Areas are Used for Cement, Tourism, Farming and Water

1) Chalk is quarried and used to make cement, which is then used to make building materials like concrete.
2) Tourists are attracted to the features of chalk and clay landscapes, e.g. escarpments and vales (see previous page).
3) There are opportunities for arable farming, livestock rearing and dairy farming on chalk and clay areas (clay vales are wide, flat, grassy areas).
4) Chalk is an aquifer — a permeable rock that holds water. Aquifers are often used as a source of drinking water — water is taken out through wells and is also pumped out of the rocks.

Case Study — The Lincolnshire Wolds, Lincolnshire

1) In 2001, 438 000 tonnes of chalk were quarried from Lincolnshire.
2) The Wolds are an Area of Outstanding Natural Beauty — many tourists come to the Wolds for the scenery and activities like walking, e.g. along the Viking Way (a long-distance footpath).
3) Around 80% of the Wolds is used as farmland to grow crops.
4) There's a major chalk aquifer underneath the Wolds — it supplies water to Lincolnshire.

Granite — bad for crops, good for stylish kitchens...

And you thought rocks were just boring grey lumps that aren't very useful. Shame on you. Learn a few of the ways granite and chalk and clay landscapes are used and maybe the rocks will forgive you. Maybe, but don't count on it.

Using Landscapes — Case Studies

I don't know about you, but I'm getting a little bit rocked out. Never mind, you're over halfway through this section now. There's just limestone area uses, a bit of tourism and some quarrying fun to go.

Limestone Areas are Used for Stone, Cement, Tourism and Farming

1) Limestone is quarried and used as a building stone to make things like floors and walls, e.g. in churches.
2) Limestone is also used to make cement.
3) Tourists are attracted to the features of limestone landscapes, e.g. limestone pavements, gorges and caverns (see page 18).
4) There are opportunities for dairy farming or rearing livestock on limestone areas. There are also opportunities for arable farming (growing crops), but in some places the soil is quite alkaline.

Case Study — The White Peak, Peak District

1) Tunstead Quarry near Buxton produces about 5.5 million tonnes of limestone every year.
2) The cement processing plant inside Tunstead Quarry makes about 800 000 tonnes of cement each year.
3) Thor's Cave is a limestone cavern in the Manifold Valley, Staffordshire — it's popular with cavers, walkers and climbers.
4) The White Peak is mainly used for intensive dairy farming.

Tourism in Any Area has Costs and Benefits

COSTS

- Large numbers of people can cause footpath erosion and littering around attractions.
- Lots of tourists cause traffic congestion.
- People may be attracted into the tourist industry from jobs like farming, causing a decline in traditional jobs.
- Many tourists own second homes in rural areas. They're absent through most of the year, so things like local shops close down.

BENEFITS

- Tourism creates jobs and brings money into the local economy.
- Farmers can diversify their business to get extra income from tourism, e.g. using buildings for camping barns or running bed and breakfasts.
- New businesses may be set up in the area to cater for the tourists, e.g. souvenir shops or hotels.

Case Study — Dartmoor, Devon

These tourists make me so congested...

COSTS

- Increased traffic means the narrow roads are getting congested. Grass verges are being damaged by tourists parking on them.
- Tourists disturb grazing animals on the moor — especially when dogs are let loose.

BENEFITS

- 4.5 million tourists visit Dartmoor every year. This creates around 3000 jobs.
- In 2003 tourism generated £120 million for the local economy.

Thor's Cave — I bet it took him ages to hammer it out...

Tourists, eh, everyone loves to hate 'em but they don't half bring a few benefits to some places. Learn about the uses of limestone areas and the benefits and costs of tourism and you win a free holiday to the White Peak*.

* Not really... sorry.

Quarrying Impacts — Case Study

People love rocks... and not just the shiny ones. No, seriously, they love them so much we have to have quarries just to meet the demand. Quarries can be quite quarrelsome things though...

Quarries Have Advantages and Disadvantages

A quarry is basically a massive pit in the ground that rock is taken from. Quarries bring advantages to some people and disadvantages to other people. This means people disagree about things like where they should go, or if we should have them at all. Here's a nice table so you can see both sides of the story:

	Advantages	Disadvantages
Economic	Quarries employ lots of local people — this brings more money into the local economy.	Tourists could be put off from visiting an area that has a quarry because they're noisy and they're an eyesore — this reduces the amount of money made from tourism.
Economic	When a quarry's built, good transport links are also built for the trucks that carry the stone — an improved infrastructure will attract other businesses and boost the local economy.	When a quarry is closed down it costs money to make them safe, e.g. by filling in holes and putting up warning signs.
Social	Some quarries are used by schools and colleges for educational visits.	People are annoyed by the heavy traffic caused by slow vehicles leaving quarries.
Social	Quarries that have been closed down can be used for recreation, e.g. for climbing.	Quarries are a very dangerous environment — people could be harmed or killed in them.
Environmental	The landscape is often restored after quarries are closed down — this can create new habitats and attract new species of wildlife to an area.	While they're operating the quarries have a massive impact on the environment — habitats are destroyed and the resources in the landscape are depleted.

Whatley Quarry is a Limestone Quarry in Somerset

Whatley Quarry is one of the largest quarries in the UK. It produces around 5 million tonnes of rock every year. Have a look at the economic, social and environmental advantages and disadvantages of Whatley Quarry:

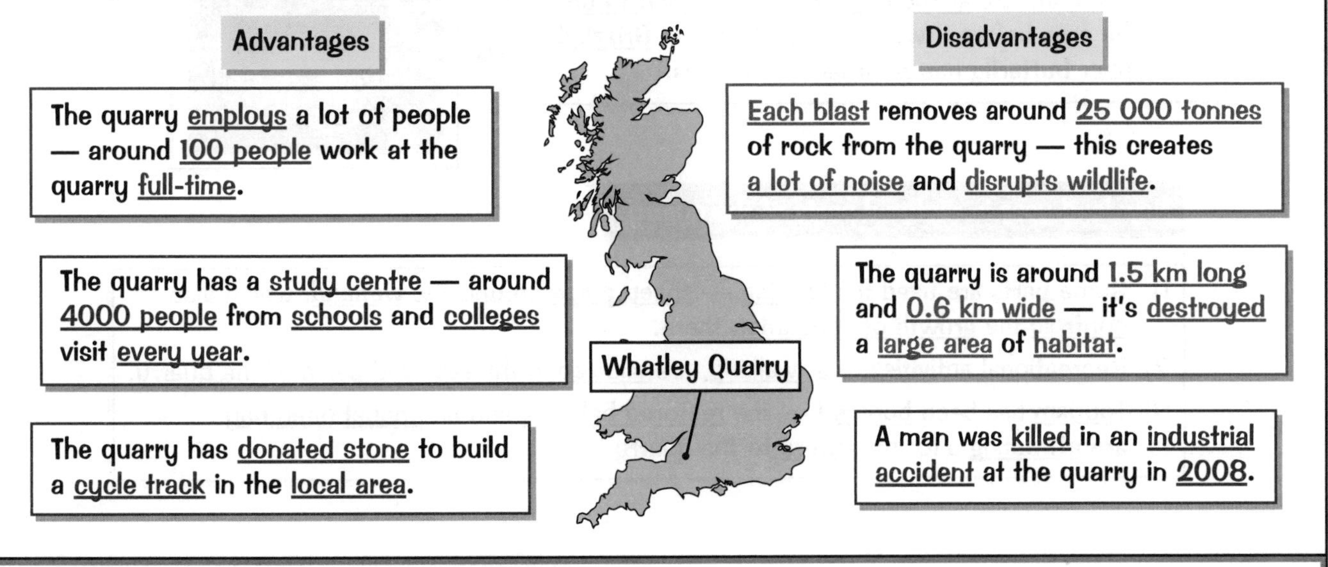

Don't dig yourself into a hole with this lot...

I'd say one of the best things about quarries is that rock gets blown up — explosions are well cool. Don't listen to me though — read all about the advantages and disadvantages of quarries, both in general and in the case study.

Quarrying Management — Case Study

My, my, aren't you lucky a person — another case study is waiting for you just below. It's another quarry, but this time it's about sustainable management... I won't keep you any longer. Have fun.

The Sustainable Management of Quarries is Important

1) Sustainable management is all about meeting the needs of people today, without hindering the ability of people in the future to meet their own needs.
2) It involves getting what we want without damaging or altering the environment in an irreversible way.
3) Quarries need sustainable management because they could seriously damage the environment, e.g. by destroying habitats and local wildlife.

Llynclys Quarry is a Limestone Quarry in Shropshire

Llynclys Quarry is a quarry in Shropshire that covers 65 hectares of land. Some quarries are abandoned when the resources are exhausted, but Llynclys Quarry is being sustainably managed in areas where extraction has finished. This minimises the environmental impact of the quarry.

Here are some of the sustainable management strategies being used:

Sustainable management strategies

1) Areas of the quarry where work has finished are being restored to the grassland, shrubland and woodland habitats that used to exist before quarrying started. About 14% of the quarry has been restored so far.

2) A wetland habitat has been created at the quarry. This encourages lots of different species to live in the area, e.g. the insects living in the wetland are a food supply for bats.

3) The habitats are attracting animals that used to be in the area before it was a quarry, e.g. the Grizzled Skipper butterfly has returned to the area.

New land uses in restored parts of the quarry

1) Some parts are used for farming — sheep graze around the wetland, which also controls the growth of vegetation there.
2) Recreational activities, e.g. walking, are allowed in the restored parts of the quarry.
3) Tourism has been boosted — the restored habitats and an annual open day are attracting a lot of visitors to the quarry.

Llynclys — send in your pronunciation suggestions on a postcard...

The hardest thing about this case study is saying the quarry's name, so you don't have any excuse for not getting to grips with the different sustainable management strategies. Now you can rock on to the revision summary...

Revision Summary for Section 2

After the blast you've had with quarrying you'd expect a revision summary to be a bit a of let down. On the contrary, this will give you a chance to shine and show the world (or the people within earshot) just how much you know about the world of rocks. If you get stuck then have a flick back through the pages, but remember — don't move on to the next section until you can answer each question with the elegance of a granite worktop.

1) a) How are igneous rocks formed?
   b) Name the two types of igneous rock.
2) What are sedimentary rocks formed from?
3) Describe how metamorphic rocks are formed.
4) What part does weathering play in the rock cycle?
5) What part does erosion play in the rock cycle?
6) Put these rock types in the order that they formed in the UK: carboniferous limestone, chalk, granite, clay.
7) What is mechanical weathering?
8) a) Describe how freeze-thaw weathering takes place.
   b) Describe how exfoliation weathering takes place.
9) What is chemical weathering?
10) Name the two types of chemical weathering.
11) Give two examples of biological weathering.
12) Describe how tors form.
13) What are moorlands?
14) How do chalk escarpments form?
15) What is a vale?
16) What is an aquifer?
17) Where does a spring line form?
18) Give two surface features of a carboniferous limestone landscape.
19) What is a resurgent river?
20) How do stalactites and stalagmites form?
21) a) Name a granite area.
    b) Give the main type of farming carried out in the area.
22) a) Name a limestone area.
    b) Give two uses of the limestone produced in the area.
23) a) Give one economic disadvantage of quarries.
    b) Give one social advantage of quarries.
24) a) Give one economic advantage of a named quarry.
    b) Give one environmental disadvantage of the same quarry.
25) a) What is sustainable management?
    b) Why is the sustainable management of quarries important?
26) a) Give one example of sustainable management during the extraction of rock at a named quarry.
    b) Give two examples of sustainable management after the extraction of rock at a named quarry.

UK Climate

You may think it rains a lot in the UK (and you'd be right)... Well, now's your chance to find out why.

The UK has a Mild Climate — Cool, Wet Winters and Warm, Wet Summers

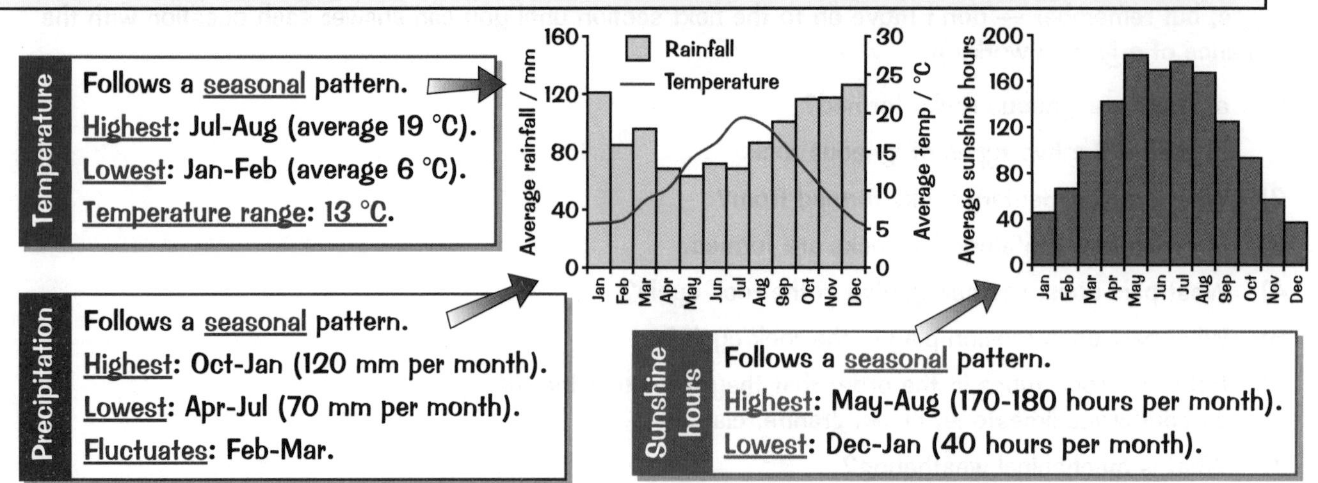

Temperature
- Follows a seasonal pattern.
- Highest: Jul-Aug (average 19 °C).
- Lowest: Jan-Feb (average 6 °C).
- Temperature range: 13 °C.

Precipitation
- Follows a seasonal pattern.
- Highest: Oct-Jan (120 mm per month).
- Lowest: Apr-Jul (70 mm per month).
- Fluctuates: Feb-Mar.

Sunshine hours
- Follows a seasonal pattern.
- Highest: May-Aug (170-180 hours per month).
- Lowest: Dec-Jan (40 hours per month).

There are Five Main Reasons Why the Climate Varies Within the UK

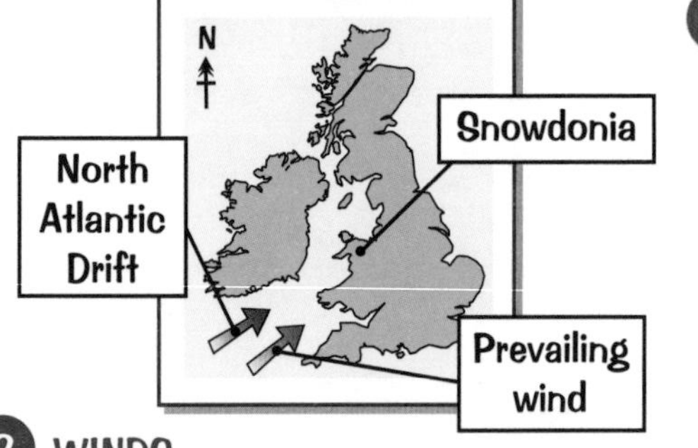

1 LATITUDE (how far north or south of the equator a place is)
- The higher in latitude you go, the colder it gets. The Sun is at a lower angle in the sky, so its heat energy is spread over more of the Earth's surface — each place receives less heat energy than at lower latitudes.
- Southern parts of the UK are warmer than northern parts because of their lower latitude.

2 WINDS
- The UK's most common (prevailing) winds are from the south west. They bring warm, moist air, which makes the UK warm and wet.
- The west of the UK gets more of the warmth and rain than the east because the winds come from the south west.

3 DISTANCE FROM THE SEA
- Areas near the sea are warmer than inland areas in winter because the sea stores up heat and warms the land.
- Areas near the sea are cooler in summer because the sea takes a long time to heat up and so cools the land down.
- The west of the UK gets warmed more than the east because of a warm ocean current coming from the south west called the North Atlantic Drift.

4 PRESSURE
- Low pressure weather systems have lots of rainfall because the air is rising and water vapour is condensing. High pressure systems have dry weather because the air is falling.
- Low pressure weather systems come from the west, so the west of the UK is wetter.

5 ALTITUDE (how high the land is)
- The higher up you go the colder it gets because the air is thinner so less heat energy is trapped.
- Higher areas get more rainfall as air is forced upwards and the water vapour condenses into rain clouds.
- So high altitude parts of the UK (e.g. Snowdonia) are colder and wetter than low altitude areas.

Some of these factors also explain why the climate of the whole of the UK is mild:
- It's not really hot or really cold because it's a mid-latitude country, and it has the North Atlantic Drift.
- The UK has both dry and rainy weather because it gets both high and low pressure weather systems.

Warm and wet — they got the wet bit right...

There's quite a lot to remember about why the climate varies in the UK, but it'll be worth your while — if you get a question about why two places in the UK have different climates you'll have this stuff nailed and you'll be laughing.

Depressions and Anticyclones

Depressions are low pressure weather systems and anticyclones are high pressure weather systems — they cause different weather. There's a depression or an anticyclone over the UK most of the time.

Depressions Form when Warm Air Meets Cold Air

Depressions form over the Atlantic ocean, then move east over the UK. Here's how they form:

1) Warm, moist air from the tropics meets cold, dry air from the poles.
2) The warm air is less dense so it rises above the cold air.
3) Condensation occurs as the warm air rises, causing rain clouds to develop.
4) Rising air also causes low pressure at the Earth's surface.
5) So winds blow into the depression in a spiral (winds always blow from areas of high pressure to areas of low pressure).
6) A warm front is the front edge of the moving warm air. A cold front is the front edge of the moving cold air.

Depressions Cause a Sequence of Weather Conditions

When a depression passes overhead you get a particular sequence of weather conditions. Imagine you're stood on the ground ahead of the warm front and the depression's moving towards you.

	5 Cold air overhead	4 As the cold front passes	3 Warm air overhead	2 As the warm front passes	1 Ahead of the warm front
Rain	Showers	Heavy showers	None	Heavy	None
Clouds	High, broken	Towering, thick	None	Low, thick	High, thin
Pressure	Rising	Suddenly rising	Steady	Falling	Falling
Temperature	Cold	Falling	Warm	Rising	Cool
Wind speed	Decreasing	Strong	Decreasing	Strong	Increasing
Wind direction	NW	SW to NW	SW	SE to SW	SE

Wind direction is given as the direction the wind comes from.

Anticyclones cause Clear Skies and Dry Weather

Anticyclones also form over the Atlantic ocean and move east over the UK. Here's a bit about them:

1) Anticyclones are where air is falling, creating high pressure and light winds blowing outwards.
2) Falling air gets warmer so no clouds are formed, giving clear skies and no rain for days or even weeks.
3) In summer, anticyclones cause long periods of hot, dry, clear weather. There are no clouds to absorb the Sun's heat energy, so more gets through to the Earth's surface causing high temperatures.
4) In winter, anticyclones give long periods of cold, foggy weather. Heat is lost from the Earth's surface at night because there are no clouds to reflect it back. The temperature drops and condensation occurs near the surface, forming fog. (It doesn't heat up much in the day because the Sun is weak.)

I don't know about the weather, but I'm depressed after that page...

Depressions have a pretty appropriate name I think. They're complicated things too, so it's really worth taking your time. Check that you can remember the weather they cause and the reasons why (N.B. it's not just because they're mean).

Extreme UK Weather

Extreme weather isn't as fun as extreme bog snorkelling, but it's becoming a lot more common in the UK.

Weather in the UK is Becoming More Extreme

1) It's raining more — the summer of 2007 was the wettest summer on record.
2) The rainfall is more intense, especially in winter — in some parts of Scotland the volume of rain that falls on wet winter days has gone up by 60%.
3) Temperature is increasing — the highest ever UK temperature was recorded in 2003 (38.5 °C).

More extreme weather has led to more extreme weather events in the last 10 years in the UK:

1) There was major flooding caused by storms and high rainfall in the south east in 2000, Cornwall in 2004, Cumbria in 2005 and 2009, the Midlands in 2007 and Devon in 2008.
2) Strong winds (combined with high tides) caused flooding in Norfolk in 2006.
3) High temperatures led to a heatwave and drought conditions in the summer of 2003.

Extreme Weather has Impacts on lots of Different Things

©istockphoto.com /Andy Green

PEOPLE'S HOMES AND LIVES

- Floods damage homes and possessions, which can cost a lot to repair or replace.
- Businesses can be damaged by floods, so people can lose their income.
- Water use can be restricted during droughts, e.g. using hosepipes can be banned.
- Increased rainfall may mean water supplies are increased.

AGRICULTURE

- Droughts can cause crop failures.
- Increased rainfall can mean higher crop yields.
- A warmer climate means farmers can grow new crops, e.g. olives.

HEALTH

- Flooding can cause deaths by drowning.
- Heatwaves can cause deaths by heat exhaustion.
- Milder winters may reduce cold-related deaths.

TRANSPORT

- Floods can block roads and railways, disrupting transport systems.
- High temperatures can cause railway lines to buckle, so trains can't run properly.

Getting to work when it floods can be a mare.

There are Three ways of Reducing the Negative Impacts

1) PREPARING — individuals and local authorities can do things to prepare for extreme weather before it happens. For example, flood defences along rivers can be improved and education programmes can tell the public the best ways to cope with floods, droughts or heatwaves.
2) PLANNING — emergency services and local councils can plan how to deal with extreme weather events in advance, e.g. they can make plans for how to rescue people from floods and where to have shelters.
3) WARNING — warning systems give people time to prepare for extreme weather. For example, the Environment Agency issues flood warnings so people can prepare and evacuate.

UK weather — the latest thing on the extreme sports channel...

Write out some of the impacts of extreme weather and the ways to reduce them. Make sure you're also clear about the evidence for the weather becoming more extreme — and looking out the window at the dreary weather doesn't count.

Global Climate Change — Debate

We British like to talk about the weather, so global climate change should give us plenty to go on...

The Earth is Getting Warmer

Climate change is any change in the weather of an area over a long period. Global warming is the increase in global temperature over the last century. Global warming is a type of climate change and it causes other types of climate change, e.g. increased rainfall. Here's a bit about the evidence for global warming:

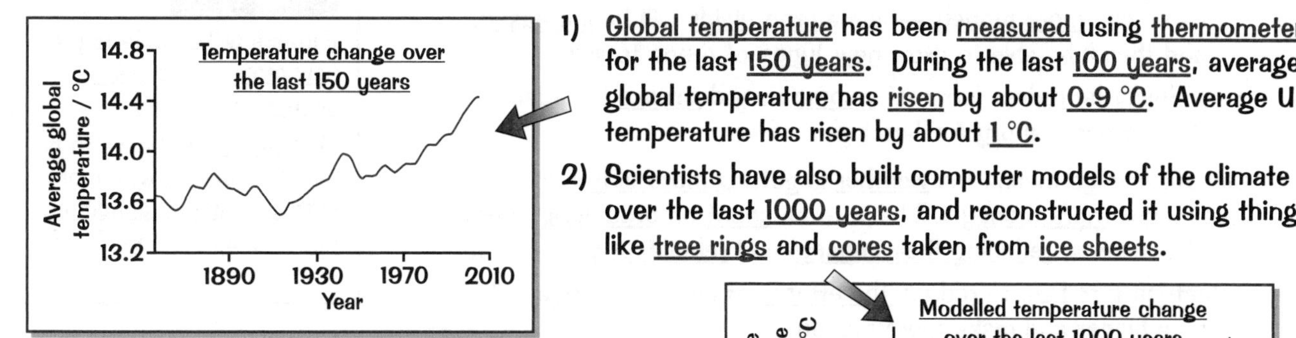

1) Global temperature has been measured using thermometers for the last 150 years. During the last 100 years, average global temperature has risen by about 0.9 °C. Average UK temperature has risen by about 1 °C.
2) Scientists have also built computer models of the climate over the last 1000 years, and reconstructed it using things like tree rings and cores taken from ice sheets.

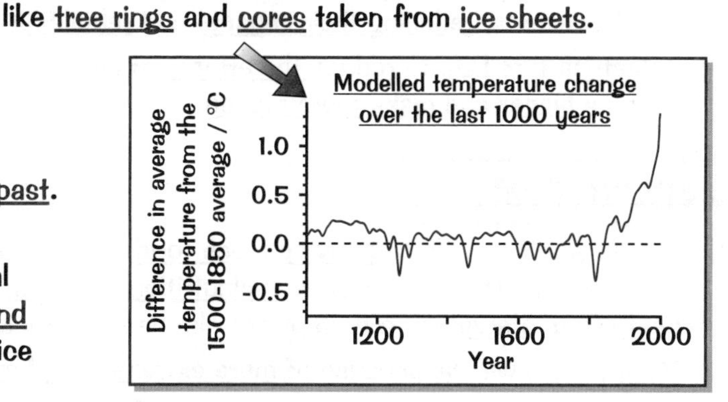

3) This shows that global temperature is rising sharply now compared to how it was in the past.
4) There's some other evidence too:
   - The ice sheets are melting because global temperature is increasing — the Greenland Ice Sheet lost an average of 195 km^3 of ice every year between 2003 and 2008.
   - Sea level is rising — increasing temperature causes ice on the land to melt and the oceans to expand. Sea level has risen 20 cm over the past century.

A few people argue that some of the evidence for global warming is a bit dodgy though. E.g. they say temperature measurements have shown an increase in temperature because human settlements have got closer to where a lot of the measurements are taken. Human settlements are warmer than natural environments because man-made surfaces like concrete absorb and radiate more heat energy.

An Increase in Greenhouse Gases is Causing Global Warming

There's a scientific consensus (general agreement) that global warming is caused by human activity:

1) An increase in human activities like burning fossil fuels, farming and deforestation has caused an increase in the concentration of carbon dioxide (CO_2) and methane (CH_4) in the atmosphere. For example, CO_2 has gone up from 280 ppm (parts per million) in 1850 to around 380 ppm today.
2) CO_2 and CH_4 are greenhouse gases — they trap heat reflected off the Earth's surface.
3) Greenhouse gases keep the Earth warm because they trap heat. Increasing the concentration of greenhouse gases in the atmosphere means the Earth heats up too much — causing global warming.

Here are some examples of other things that can cause climate change:

1) Variations in solar output — the Sun's output of energy isn't constant. In periods when there's more energy coming from the Sun, the Earth gets warmer.
2) Changes in the Earth's orbit — the way the Earth orbits the Sun changes, which affects how much energy the Earth receives. If the Earth receives more energy it gets warmer.

Which climate shall I change into today — the hot or the cold one...

The climate is changing — global warming is happening, it's just that a handful of people think some of the evidence isn't great. There are other things that cause climate change, but let's face it, we humans better take the rap this time.

Global Climate Change — Impacts

Global climate change will have economic, social, environmental and political impacts on the world and on the UK. Boy, that's quite a few impacts, but then, the climate's kind of important you know.

Climate Change will have Economic Impacts...

1) Climate change will affect farming in different ways around the world:
   - In higher latitudes, warmer weather will mean some farmers can make more money — some crop yields will be increased, and they'll be able to grow new types of crops to sell.
   - In lower latitudes, farmers' income may decrease because it's too hot and dry for farming.
2) Climate change means the weather is getting more extreme. This means more money will have to be spent on predicting extreme weather events, reducing their impacts and rebuilding after them.
3) Industries that help to reduce the effects of climate change will become bigger and make more money.

IN THE UK...

Farmers will be able to grow new crops in the warmer climate, e.g. olives.

More money will have to be spent on coping with more extreme weather conditions, e.g. to pay for more flood defences.

...Social Impacts...

1) People won't be able to grow as much food in lower latitudes (see above). This could lead to malnutrition, ill health and death from starvation, e.g. in places like central Africa.
2) More people will die because of more extreme weather events.
3) Hotter weather makes it easier for some infectious diseases to spread. This will lead to more ill health and more deaths from disease.
4) Some areas will become so hot and dry that they're uninhabitable. People will have to move, which could lead to overcrowding in other areas.

IN THE UK...

There could be fewer cold-related deaths, but more deaths caused by hot weather, e.g. from heat exhaustion.

Diseases that don't exist in the UK at the moment could become common, e.g. malaria.

Don't forget — global warming is a type of climate change.

...Environmental Impacts...

1) Global warming is causing sea level to rise, so some habitats will be lost as low-lying coastal environments are submerged.
2) Rising temperature and decreased rainfall will mean some environments will turn into deserts.
3) The distribution of some species may change due to climate change (species can only live in the areas where the conditions suit them best). Species that can't move may die out.

IN THE UK...

Flooding and sea level rise is threatening some coastal habitats, e.g. in the south east and Norfolk.

The distribution of some species in the UK may change, e.g. it's thought beech trees will become more common in Scotland.

...and Political Impacts... phew

1) Water will become more scarce in some places. Competition over water could lead to war between countries.
2) Climate change may cause people to move (see above). This means some countries will have to cope with increased immigration and emigration.
3) Governments are under pressure to come up with ways to slow climate change or reduce its effects.

IN THE UK...

The government has had to set up a new political department to come up with ways to slow climate change and reduce its impacts — the Department for Energy and Climate Change.

Olives in the garden and warmer weather — bring on global warming...

The UK hasn't managed to escape climate change, so you should check that you know the impacts on the world and the UK too. The impact on me right now is that I'm roasting hot and could do with a nice cool glass of fruit juice.

Global Climate Change — Responses

Most of the responses to climate change involve cutting emissions of greenhouse gases like CO_2. This can be done globally, nationally and locally, so everyone gets a slice of the fun.

The Kyoto Protocol was a Global Response

The Kyoto Protocol was due to expire at the end of 2012, but many countries agreed to extend it to 2020.

From 1997, most countries in the world agreed to monitor and cut greenhouse gas emissions by signing an international agreement called the Kyoto Protocol:

1) The aim was to reduce global greenhouse gas emissions by 5% below 1990 levels by 2012.
2) Each country was set a target, e.g. the UK agreed to reduce emissions by 12.5% by 2012.
3) Another part of the protocol was the carbon credits trading scheme:

- Countries that came under their emissions target got carbon credits which they could sell to countries that didn't meet their emissions target. This meant there was a reward for having low emissions.
- Countries could also earn carbon credits by helping poorer countries to reduce their emissions. The idea was that poorer countries would be able to reduce their emissions more quickly.

4) Not all countries agreed to the Kyoto Protocol though — the USA didn't agree, and they have the highest emissions of any country in the world (22% of global CO_2 emissions in 2004).

There are also National and Local Responses to Climate Change

NATIONAL RESPONSES

1) **TRANSPORT STRATEGIES**
Governments can improve public transport networks like buses and trains. For example, they can make them run faster or cover a wider area. This encourages more people to use public transport instead of cars, so CO_2 emissions are reduced.

2) **TAXATION**
Governments can increase taxes on cars with high emissions, e.g. in the UK there are higher tax rates for cars with higher emissions. This encourages people to buy cars with low emissions, so emissions are reduced.

LOCAL RESPONSES

1) **CONGESTION CHARGING**
Local authorities can charge people for driving cars into cities during busy periods, e.g. there's a congestion charge to drive into central London during busy times of the day. This encourages people to use their cars less, which reduces emissions.

2) **RECYCLING**
- Local authorities can recycle more waste by building recycling plants and giving people recycling bins. Recycling materials means less energy is used making new materials, so emissions are reduced.
- Local authorities can also create energy by burning recycled waste, e.g. Sheffield uses a waste incinerator to supply 140 buildings with energy.

3) **CONSERVING ENERGY**
- Local authorities give money and advice to make homes more energy efficient, e.g. by doing things like improving insulation. This means people use less energy to heat their homes, because less is lost. Emissions are reduced because less energy needs to be produced.
- Individuals can also conserve energy by doing things like switching lights off and not leaving electric gadgets on standby.

My response to climate change — slap on the suncream and bust out the shades...

Climate change is a global problem, so the response to deal with it needs to be on a global scale. That means everyone has to do their bit, from world leaders down to folk like us. Now, did I leave my hair straighteners on...

Tropical Storms

Tropical storms are intense low pressure weather systems. They've got lots of different names (hurricanes, typhoons, tropical cyclones, tropical revolving storms and willy willies), but they're all the same thing.

Tropical Storms Develop over Warm Water

Tropical storms are huge storms with strong winds and torrential rain. Scientists don't know exactly how they're formed, but they know where they form and some of the conditions that are needed:

1) Tropical storms develop above sea water that's 27 °C or higher.
2) They happen when sea temperatures are highest, so they happen at different times in different places. For example, tropical storm season is August to October in the Atlantic, and May to December in the north east Pacific.
3) Warm, moist air rises and condensation occurs. This releases huge amounts of energy, which makes the storms really powerful.
4) They move west because of the easterly winds near the equator.

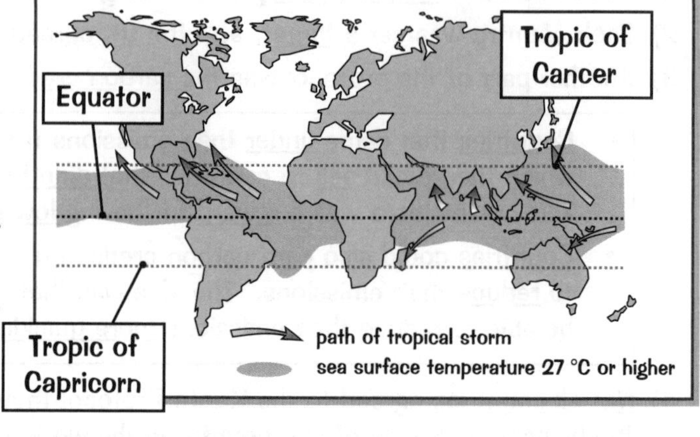

5) They lose strength as they move over land because the energy supply from the warm water is cut off.
6) Most tropical storms occur between 5° and 30° north and south of the equator, e.g. in the Atlantic and the Indian Ocean (any further from the equator and the water isn't warm enough).
7) The Earth's rotation deflects the path of the winds, which causes the storms to spin.

Tropical Storms are Circular from Above

1) Tropical storms spin anticlockwise and move north west (in the northern hemisphere).
2) They're circular in shape and can be hundreds of kilometres wide.
3) They usually last between 7 and 14 days.
4) The centre of the storm's called the eye — it's up to 50 km across and is caused by descending air. There's very low pressure, light winds, no clouds and no rain in the eye.

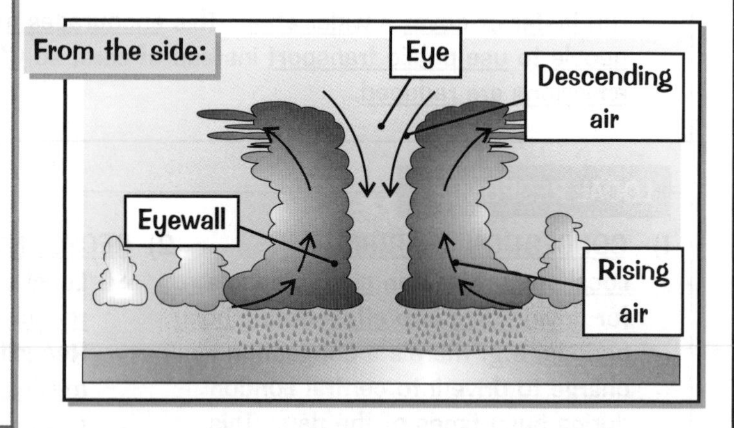

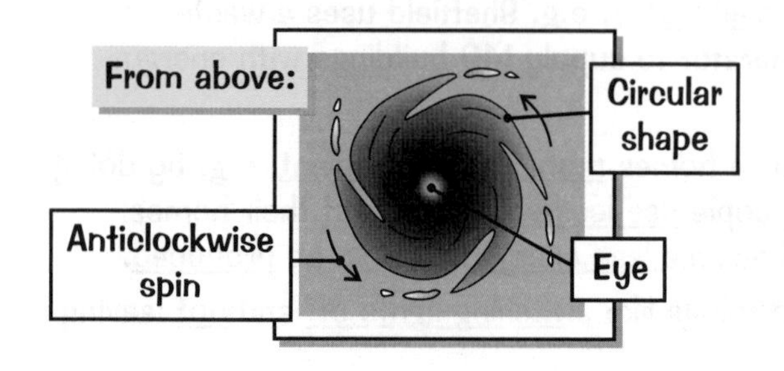

5) The eye is surrounded by the eyewall, where there's spiralling rising air, very strong winds (around 160 km per hour), storm clouds and torrential rain.
6) Towards the edges of the storm the wind speed falls, the clouds become smaller and more scattered, and the rain becomes less intense.

Forget warm water, you're in hot water when one of these turns up...

Well, that blew the cobwebs away. Since all the top scientists haven't worked it out yet, you don't need to know exactly how a tropical storm forms, but check you know where they're found, why they're found there and their characteristics.

Impacts of Tropical Storms

Tropical storms try really hard to make an impact. Unfortunately, no one ever appreciates their effort...

Tropical Storms have Primary and Secondary Impacts

The primary impacts of a tropical storm are the immediate effects of strong winds, high rainfall and storm surges. The secondary impacts are the effects that happen later on. Here are a few examples of the possible impacts:

Storm surges are large rises in sea level caused by the low pressure and high winds of a storm.

Primary impacts

1) Buildings and bridges are destroyed.
2) Rivers and coastal areas flood.
3) People drown, or they're injured or killed by debris that's blown around.
4) Roads, railways, ports and airports are damaged.
5) Electricity cables are damaged, cutting off supplies.
6) Telephone poles and cables are destroyed.
7) Sewage overflows due to flooding. The sewage often contaminates water supplies.
8) Crops are damaged and livestock is killed.
9) Heavy rain makes hills unstable, causing landslides.
10) Beaches are eroded and coastal habitats (e.g. coral reefs) are damaged.

Damage from Hurricane Katrina, 29th August 2005.

Secondary impacts

1) People are left homeless.
2) There's a shortage of clean water and a lack of proper sanitation — this makes it easier for diseases to spread.
3) Roads are blocked or destroyed so aid and emergency vehicles can't get through.
4) Businesses are damaged or destroyed, causing unemployment.
5) There's a shortage of food because crops are damaged and livestock has died.
6) People may suffer psychological problems if they knew people who died.

The more settlements built and businesses set up in an area, the greater the impact because there are more people and properties to be affected by a tropical storm.

The Impacts of Tropical Storms are More Severe in Poorer Countries

Here are a few reasons why:

1) There's more low quality housing in poorer countries. Low quality houses are destroyed more easily by strong winds and flooding.
2) The infrastructure is often worse in poorer countries. Poor quality roads make it harder for emergency services to rescue people, which leads to more deaths.
3) More people in poorer countries depend on farming. If crops and livestock are destroyed lots of people will lose their livelihoods, and some might starve.
4) Poorer countries don't have much money to protect against tropical storms, e.g. by building flood defences. They also don't have enough money or resources (e.g. food and helicopters) to react straight away to tropical storms, so more people are affected by secondary impacts.
5) Healthcare is often worse in poorer countries. Many of the hospitals don't have enough supplies to deal with the large numbers of casualties after a tropical storm, so more people die from treatable injuries.

People Continue to Live in the Areas where Tropical Storms Happen

The reasons why people don't move away from areas that are prone to tropical storms are the same as why people don't move away from areas prone to earthquakes (see page 5) — they don't want to leave friends, they've got a job in the area, they don't think a tropical storm will happen again so it's safe etc.

Getting smacked in the face by a wet fish — a rare impact of tropical storms...

Tropical storms pretty much wreak havoc in the areas they hit — boats get tossed into trees, hills go for a slide and the winds are so strong you can't stand upright... you want to get out of there faster than Superman in a kryptonite mine.

Reducing the Impacts of Tropical Storms

I think I'm psychic — I just knew this page was going to be about reducing the impacts of tropical storms.

There are Many Ways of Reducing the Impacts of Tropical Storms

Prediction

©iStockphoto.com

1) When and where tropical storms will hit land can be predicted. Scientists use data from things like radar, satellites and aircraft to track the storm. Computer models are then used to calculate a predicted path for the storm.
2) Predicting where and when a tropical storm is going to happen gives people time to evacuate — this reduces the number of injuries and deaths. It also gives them time to protect their homes and businesses, e.g. by boarding up windows so they don't get smashed.

Planning

1) Future developments, e.g. new houses, can be planned to avoid the areas most at risk (e.g. right on the coast). This reduces the number of buildings destroyed by winds or flooding.
2) Emergency services can train and prepare for disasters, e.g. by practising rescuing people from flooded areas with helicopters. This reduces the number of people killed.
3) Governments can plan evacuation routes to get people away from storms quickly. This reduces the number of people injured or killed by things like flying debris or floodwater.

A hurricane evacuation route sign in Florida, USA.

Building techniques

1) Buildings can be designed to withstand tropical storms, e.g. by using reinforced concrete or by fixing roofs securely so they don't get blown off. Buildings can also be put on stilts so they're safe from floodwater.
2) Flood defences can be built along rivers (e.g. levees) and coasts (e.g. sea walls).
3) All of these reduce the number of buildings destroyed, so fewer people will be killed, injured, made homeless and made unemployed.

Education

1) Governments and other organisations can educate people about how to prepare for a tropical storm (e.g. by stockpiling water and food) and how to evacuate. This helps reduce deaths.
2) People can be told how to make a survival kit containing things like food, water and medication. The kits reduce the chance of people dying if they're stuck in the area.

Aid

Governments or organisations often send aid to countries hit by tropical storms, e.g. food, bottled water, tents. This helps to reduce the impacts, e.g. food aid stops people going hungry.

Some Strategies are More Sustainable than Others

There's a definition of a sustainable strategy way back on page 6.
Here's a bit on the sustainability of strategies to reduce the impact of tropical storms:

1) All of the strategies are sustainable because they're all effective and environmentally friendly.
2) Some are more cost-effective than others though, so are more sustainable.
3) Predicting tropical storms needs special equipment (e.g. radars) and trained scientists, which makes it expensive, but if it's accurate it saves a lot of lives.
4) Building techniques can be very expensive, but can save a lot of money if they stop building destruction.

I predict you'll need aid if you don't get your head around this page...

Prediction means you know if you have to do some serious plywood DIY and hightail it out of there. Good evacuation planning means you can hightail quickly and safely. Other things help make sure you've got a house to come back to.

Tropical Storms — Case Studies

Tropical storms can wreak quite a lot of havoc you know. Here are a couple of case studies...

Tropical Storms have Different Effects in Different Places

The effects of tropical storms and the responses to them are different in different parts of the world. A lot depends on how wealthy the part of the world is. Here are a couple of case studies, so you can cash in those marks when you get asked to compare two case studies in the exam.

Tropical storm in a rich part of the world:

Name: Hurricane Katrina
Place: South east USA
Date: 29th August, 2005

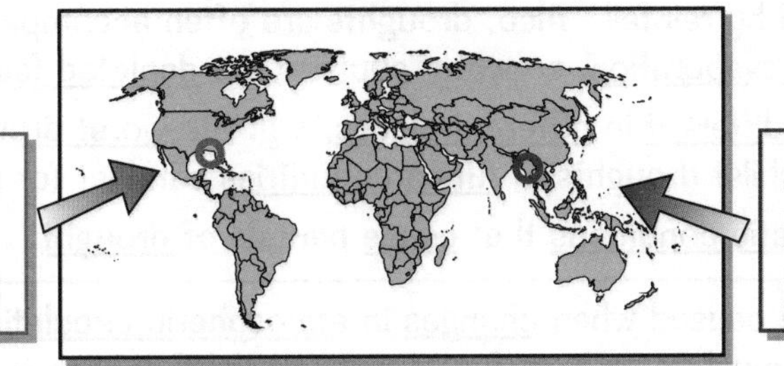

Tropical storm in a poor part of the world:

Name: Cyclone Nargis
Place: Irrawaddy delta, Burma
Date: 2nd May, 2008

	Hurricane Katrina (USA)	Cyclone Nargis (Burma)
Preparation	• The USA has a sophisticated monitoring system to predict if a hurricane will hit (e.g. by using satellite images of the Atlantic) — so people were warned. • Mississippi and Louisiana declared states of emergency on 26th August — they set up control centres and stockpiled supplies. • 70-80% of New Orleans residents were evacuated before the hurricane reached land.	• Indian and Thai weather agencies warned the Burmese Government that Cyclone Nargis was likely to hit the country. Despite this, Burmese forecasters reported there was little or no risk. • There were no emergency or evacuation plans.
Social effects	• More than 1800 people were killed. • 300 000 houses were destroyed. • 3 million people left without electricity. • One of the main routes out of New Orleans was closed because parts of the I-10 bridge collapsed.	• More than 140 000 people were killed. • 450 000 houses were destroyed. • 2-3 million people were made homeless. • 1700 schools were destroyed.
Economic effects	• Total of around $300 billion of damage. • 230 000 of jobs were lost from businesses that were damaged or destroyed. • 30 offshore oil platforms sunk or went missing. This increased the price of fuel. • Shops in New Orleans were looted by residents in the days after the hurricane.	• Total of around $4 billion of damage. • Millions of people lost their livelihoods. • 200 000 farm animals were killed, crops were lost and over 40% of food stores were destroyed.
Environmental effects	• The hurricane caused the sea to flood parts of the land. This destroyed some coastal habitats, e.g. sea turtle breeding beaches.	• Coastal habitats such as mangrove forests were damaged. • The salinity (salt content) of soil in some areas has increased because of flooding by sea water. This means it's more difficult for plants to grow.
Short-term response	• During the storm the coast guard, police, fire service, army and volunteers rescued over 50 000 people. • About 25 000 people were given temporary shelter at a sports stadium (the Louisiana Superdome) immediately after the storm.	• Burma's Government initially refused to accept any foreign aid. Aid workers were only allowed in 3 weeks after the disaster occurred. • The UN launched a massive appeal to raise money to help respond to the disaster.
Long-term response	• The US government has spent over $800 million on rebuilding flood defences. • Around $34 billion has been set aside for the re-building of things like houses and schools.	• Burma is relying on international aid to repair the damage — fewer than 20 000 homes have been rebuilt and half a million survivors are still living in temporary shelters.

Katrina — sounds like a nice girl...

When you compare the two, it's clear that the impacts were worse in Burma and the long-term response has been a bit more organised in the USA. Get both case studies etched into your memory and you'll be ready for anything.

Drought

Blimey, I bet you're wondering when all this weather and climate malarkey is going to end. Well, you're in luck — the answer is 'soon'. Yes indeedy, there are just three lovely pages on drought left to go...

Drought is when Conditions are Drier than Normal

1) A drought is a long period (weeks, months or years) when rainfall is below average.
2) Water supplies, e.g. lakes and rivers, are depleted during a drought because people keep using them but they aren't replenished by rainfall. Also, droughts are often accompanied by high temperatures, which increase the rate of evaporation, so water supplies are depleted faster.
3) The length of a drought is different in different places, e.g. the worst drought in Britain since records began lasted 16 months, whilst droughts in African countries can last for more than a decade.
4) Here are a few of the climatic conditions that cause periods of drought:

> 1) Droughts are caused when changes in atmospheric circulation mean it doesn't rain much in an area for years, e.g. this happens in Ethiopia.
> 2) Changes in atmospheric circulation can also make the annual rains fail (e.g. monsoon rains don't come when they normally do in places like India).
> 3) Droughts are also caused when high pressure weather systems (called anticyclones) block depressions (weather systems that cause rain), e.g. this happens in the UK.

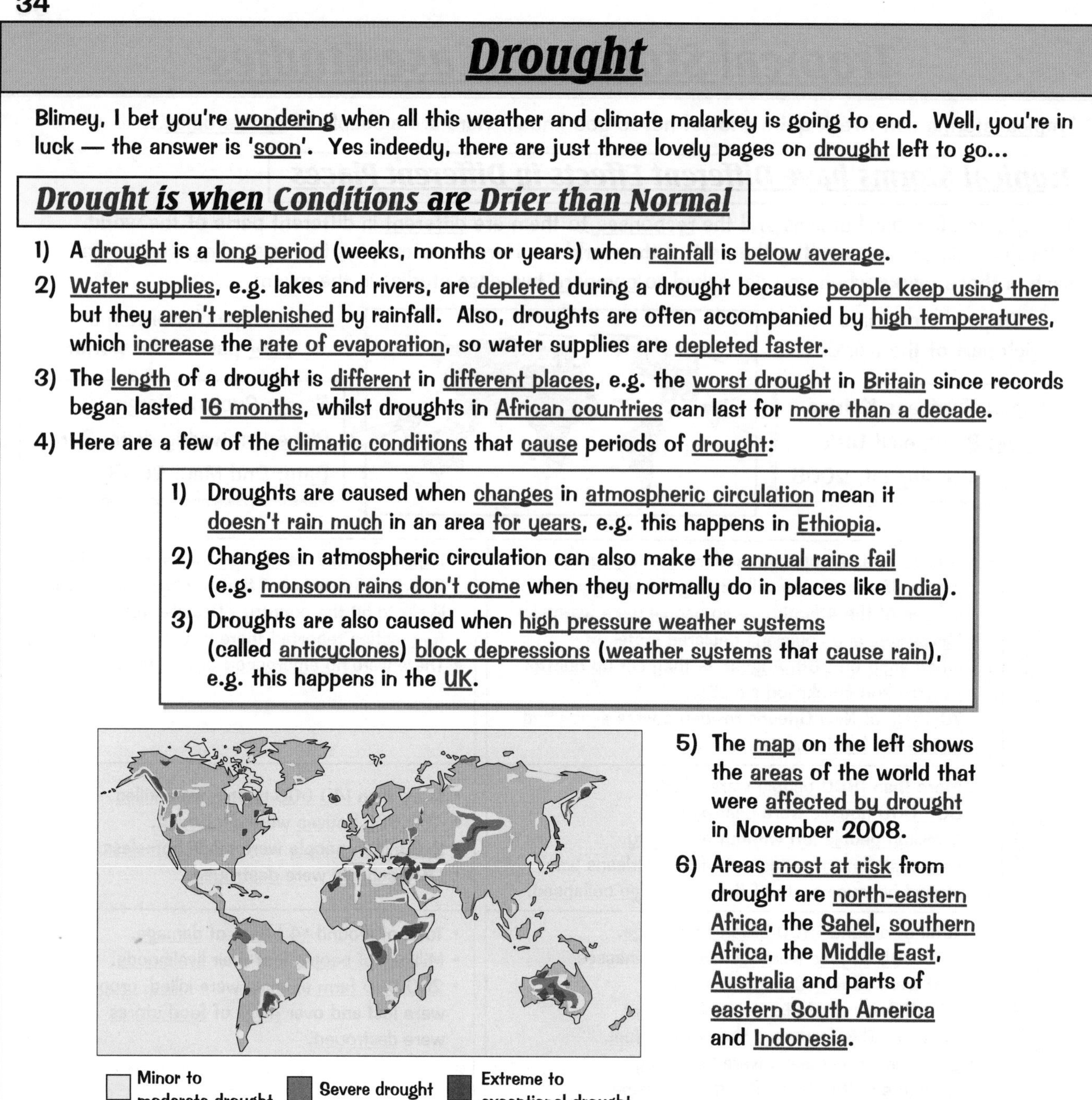

5) The map on the left shows the areas of the world that were affected by drought in November 2008.
6) Areas most at risk from drought are north-eastern Africa, the Sahel, southern Africa, the Middle East, Australia and parts of eastern South America and Indonesia.

People Continue to Live in the Areas where Droughts Happen

The reasons why people don't move away from areas that are prone to droughts are similar to why people don't move away from areas prone to earthquakes (see page 5) — they've always lived there, they've got a job in the area, they don't think a drought will happen again soon etc.

©iStockphoto.com/Klaas Lingbeek-van Kranen

Droughts and Tropical Storms are Climatic Hazards

1) Climatic hazards are natural hazards caused by the weather.
2) A natural hazard is a naturally occurring event that has the potential to affect people's lives or property.

Droughts are quite sarcastic — they have a dry sense of humour...

Now you know what droughts are and how they happen. If you don't, then you've come to the wrong place — this book is most useful if you work your way down pages, from top to bottom. It's an easy mistake to make though...

Impacts of Droughts

I know what you're thinking — 'please, no more hazard impacts'. Your wish is my command (after this page).

Droughts have Primary and Secondary Impacts

The primary impacts of droughts are the immediate effects of low rainfall and reduced water supplies. Secondary impacts are the effects that happen later on. Here are some examples of the possible impacts:

Primary impacts

1) Vegetation dies (including crops).
2) People and animals die from dehydration.
3) Aquatic animals (e.g. fish) die because lakes and rivers dry up.
4) Soil dries out and is easily eroded by the wind and rain.

©iStockphoto.com/Morley Read

Secondary impacts

1) Animals die from starvation because there's no vegetation.
2) There's a shortage of food because crops have failed and livestock has died, so people die from starvation.
3) Soil erosion is increased because there's less vegetation to hold it together. This causes desertification — where land becomes unsuitable for growing vegetation.
4) There are conflicts over water supplies.
5) People move out of the area to find water.
6) Farms close, causing unemployment.
7) People may suffer psychological problems, e.g. stress from losing their business.
8) Dried out vegetation can be easily ignited, e.g. by lightning, causing wildfires.
9) Winds pick up dry soil, causing dust storms.

Some Human Activities Increase the Impacts of Droughts

Some of the things that humans do make the impacts of droughts worse. Here are a couple of examples:

Overgrazing

Overgrazing reduces vegetation in an area. This makes the soil erosion caused by droughts even worse — with fewer plants, soil isn't held together as strongly so it's eroded more easily.

As usual, the more settlements built and farms set up in an area, the greater the impacts because there are more people and businesses to be affected by drought.

Excessive irrigation

Irrigation is where water is artificially supplied from rivers or lakes to farmland to increase crop production. However, excessive irrigation depletes rivers and lakes, which increases the impact of drought because there's less water. Also, when irrigation water evaporates, salts are left in the soil (this is salinisation). Crops don't grow well in salty soil, so this also increases the impact of drought.

The Impacts of Droughts are More Severe in Poorer Countries

Droughts happen all over the world, but they have a greater impact in poorer countries. Here's why:

1) More people in poorer countries depend on farming. If crops and livestock die lots of people will lose their livelihoods and some might starve.
2) Poorer countries have less money to prepare for droughts or respond to them, e.g. they can't afford to build reservoirs (see the next page), so the impacts of a drought are more severe.

Make no bones about it — the impacts of drought aren't that great...

The stuff on this page isn't too tricky so it'll be nice and easy to store it in your memory box. Then you'll know all about the two human activities that make the impacts of drought a lot worse. Silly humans, when will they learn...

Reducing the Impacts of Droughts

Luckily there are ways to reduce the impacts of droughts. Some of them are more sophisticated than my rain dance (a cross between the moonwalk and the funky chicken) and surprisingly they work much better.

There are Many Ways of Reducing the Impacts of Droughts

Prediction

1) Droughts can be predicted a short time before they happen by monitoring rainfall, soil moisture and river levels.
2) When a drought is predicted, things can be done to reduce the impacts, e.g. banning hosepipes, rationing water or moving people out of areas that will be worst affected.

Farming techniques

1) Drought-resistant crops (ones that need little water) can be grown, e.g. millet, sorghum and olives.
2) More efficient methods of irrigation can be used. For example, drip irrigation delivers small volumes of water directly to crop roots (reducing the amount lost by evaporation).
3) These techniques reduce the demand on water supplies and make food production more reliable.

©iStockphoto.com/Lucas Cavalheiro

Drip irrigation

Water conservation

1) People can conserve water by reducing the amount they use in their homes, e.g. by installing low volume flush toilets, and by having showers instead of baths.
2) People can also install water butts at home to collect rainwater and use it to wash their car or water their garden.
3) These reduce the demand on water supplies, so more water is available during a drought.

Increase water supplies

1) Reservoirs and wells can be built to increase water supplies.
2) These make more water available during a drought, reducing deaths from dehydration, reducing conflicts over supplies and making food production more reliable.

Reservoirs are man-made lakes that store water — they're created behind dams that are built across rivers.

A rural well

Aid

Aid can help reduce the impacts of drought in more than one way:

- Emergency aid (like food and water tankers) can stop people dying from dehydration or starvation.
- Aid can be used to fund development projects, e.g. building wells or water pipes, to make more water available during droughts.

Some Strategies are More Sustainable Than Others

There's a definition of a sustainable strategy back on page 6. Here's a bit about the sustainability of strategies to reduce the impact of drought:

1) Most of the strategies are sustainable because they're effective and environmentally friendly. Building wells can deplete groundwater supplies, which means there's less water for people in the future, so sometimes they're not sustainable. Building reservoirs can reduce other people's water supply downriver. This means it's not sustainable as it doesn't meet the needs of people alive now.
2) As usual, some are more sustainable than others because they're more cost-effective, e.g. buying pipes for drip irrigation can be expensive, but it saves a lot of water and can be more cost-effective than emergency aid.

That's it, my humour supply is depleted — it wasn't even that full to begin with...

I hope this section has quenched your thirst for knowledge about weather and climate because I've got nothing else to tell you. But hold on — the forecast for the next page has predicted a heavy downpour... of questions. I am hilarious.

Revision Summary for Section 3

Well, wasn't that a blast of fresh air from the prevailing wind. Weather and climate is a pretty complicated section, so don't worry if it didn't sink in first time round. Give these questions a bit of a whirl to see whether your weather knowledge is up to the task. If you're still in a hurricane of confusion, then look back through the section and give the bits you don't know the once over — and don't let the depressions get you down.

1) During what months are temperatures highest in the UK?
2) Explain why latitude causes the climate within the UK to vary.
3) Give one reason why the whole of the UK has a mild climate.
4) Fill in the blanks below:
   Depressions form when _____ _____ air from the tropics, meets _____ _____ air from the poles.
5) Describe the weather conditions as a cold front passes overhead.
6) What weather conditions do anticyclones bring in winter?
7) Give one piece of evidence for the weather becoming more extreme in the UK.
8) Give two impacts that extreme weather has on the homes and lives of people in the UK.
9) Describe one way that the negative impacts of extreme weather in the UK are being reduced.
10) What is global warming?
11) Give one piece of evidence for global warming.
12) How do greenhouse gases cause global warming?
13) Give one economic impact and one environmental impact of climate change on the world.
14) Give one social impact and one political impact of climate change in the UK.
15) How much did the Kyoto Protocol aim to reduce global greenhouse gas emissions by?
16) Describe one national response to climate change.
17) Give one condition that's needed for tropical storms to form.
18) Describe the distribution of tropical storms.
19) In what part of a tropical storm are there no clouds overhead?
20) Describe four secondary impacts of tropical storms.
21) Give three reasons why the impacts of tropical storms are more severe in poorer countries.
22) Describe two ways buildings can be designed to withstand a tropical storm.
23) Why is it expensive to predict tropical storms?
24) a) Name one tropical storm that happened in a rich part of the world.
    b) Give three effects of this tropical storm.
    c) Give one long-term response to this tropical storm.
25) What is a drought?
26) Describe one climatic condition that causes droughts to happen.
27) Give one reason why people continue to live in the areas where droughts happen.
28) What is a climatic hazard?
29) Describe five secondary impacts of a drought.
30) How does overgrazing increase the impacts of droughts?
31) Describe one reason why the impacts of droughts are more severe in poorer countries.
32) Give two examples of how people can conserve water at home to reduce the impacts of droughts.
33) Explain how aid can be used to help reduce the impacts of droughts.

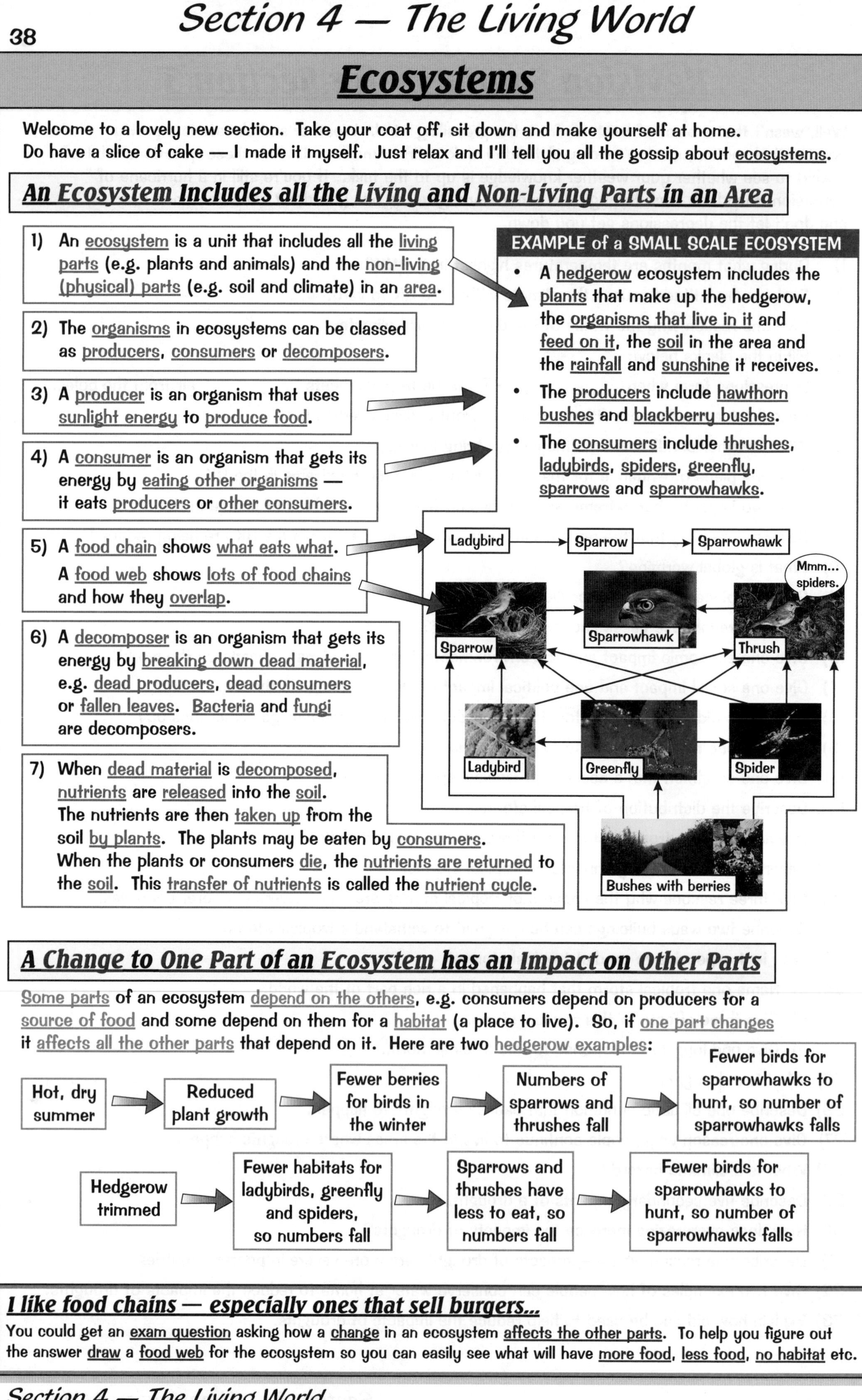

Ecosystems

Welcome to a lovely new section. Take your coat off, sit down and make yourself at home. Do have a slice of cake — I made it myself. Just relax and I'll tell you all the gossip about ecosystems.

An Ecosystem Includes all the Living and Non-Living Parts in an Area

1) An ecosystem is a unit that includes all the living parts (e.g. plants and animals) and the non-living (physical) parts (e.g. soil and climate) in an area.
2) The organisms in ecosystems can be classed as producers, consumers or decomposers.
3) A producer is an organism that uses sunlight energy to produce food.
4) A consumer is an organism that gets its energy by eating other organisms — it eats producers or other consumers.
5) A food chain shows what eats what. A food web shows lots of food chains and how they overlap.
6) A decomposer is an organism that gets its energy by breaking down dead material, e.g. dead producers, dead consumers or fallen leaves. Bacteria and fungi are decomposers.
7) When dead material is decomposed, nutrients are released into the soil. The nutrients are then taken up from the soil by plants. The plants may be eaten by consumers. When the plants or consumers die, the nutrients are returned to the soil. This transfer of nutrients is called the nutrient cycle.

EXAMPLE of a SMALL SCALE ECOSYSTEM

- A hedgerow ecosystem includes the plants that make up the hedgerow, the organisms that live in it and feed on it, the soil in the area and the rainfall and sunshine it receives.
- The producers include hawthorn bushes and blackberry bushes.
- The consumers include thrushes, ladybirds, spiders, greenfly, sparrows and sparrowhawks.

A Change to One Part of an Ecosystem has an Impact on Other Parts

Some parts of an ecosystem depend on the others, e.g. consumers depend on producers for a source of food and some depend on them for a habitat (a place to live). So, if one part changes it affects all the other parts that depend on it. Here are two hedgerow examples:

Hot, dry summer → Reduced plant growth → Fewer berries for birds in the winter → Numbers of sparrows and thrushes fall → Fewer birds for sparrowhawks to hunt, so number of sparrowhawks falls

Hedgerow trimmed → Fewer habitats for ladybirds, greenfly and spiders, so numbers fall → Sparrows and thrushes have less to eat, so numbers fall → Fewer birds for sparrowhawks to hunt, so number of sparrowhawks falls

I like food chains — especially ones that sell burgers...

You could get an exam question asking how a change in an ecosystem affects the other parts. To help you figure out the answer draw a food web for the ecosystem so you can easily see what will have more food, less food, no habitat etc.

World Ecosystems

I hope you've packed suitable clothes because I'm about to take you on a whistle-stop tour of the world's ecosystems. Well, we'll go to some of them anyway (really cold ones are out because I can't find my hat).

Different Parts of the World Have Different Ecosystems

1) The climate in an area determines what type of ecosystem forms. So different parts of the world have different ecosystems because they have different climates.
2) The map shows the global distribution of three types of ecosystem.
3) Tropical rainforests are found around the equator.
4) Hot deserts are found between 15° and 30° north and south of the equator where there's less rainfall.
5) Temperate deciduous forests are found between 40° and 60° north and south of the equator in places where there are four distinct seasons.

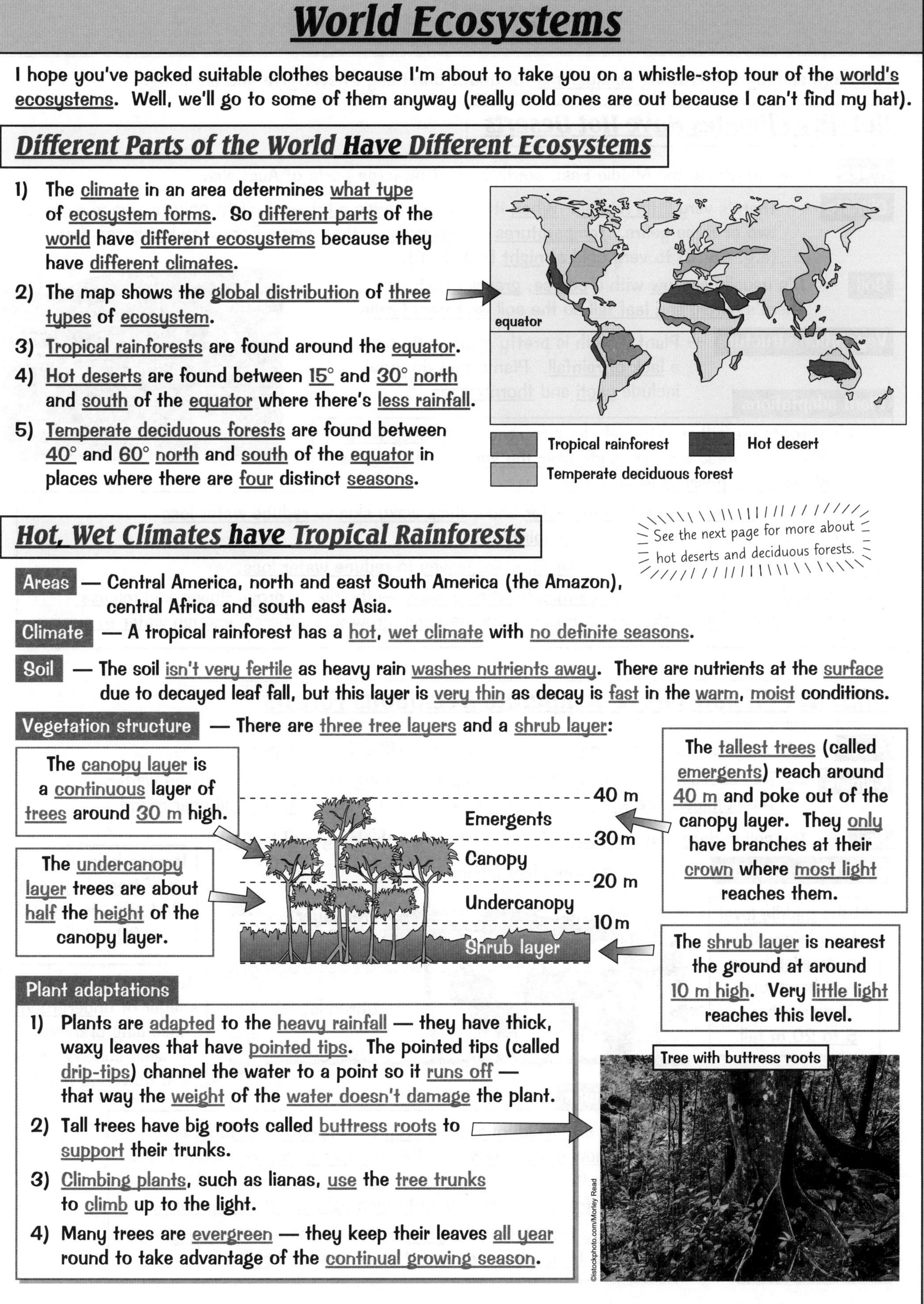

See the next page for more about hot deserts and deciduous forests.

Hot, Wet Climates have Tropical Rainforests

Areas — Central America, north and east South America (the Amazon), central Africa and south east Asia.

Climate — A tropical rainforest has a hot, wet climate with no definite seasons.

Soil — The soil isn't very fertile as heavy rain washes nutrients away. There are nutrients at the surface due to decayed leaf fall, but this layer is very thin as decay is fast in the warm, moist conditions.

Vegetation structure — There are three tree layers and a shrub layer:

The canopy layer is a continuous layer of trees around 30 m high.

The undercanopy layer trees are about half the height of the canopy layer.

The tallest trees (called emergents) reach around 40 m and poke out of the canopy layer. They only have branches at their crown where most light reaches them.

The shrub layer is nearest the ground at around 10 m high. Very little light reaches this level.

Plant adaptations

1) Plants are adapted to the heavy rainfall — they have thick, waxy leaves that have pointed tips. The pointed tips (called drip-tips) channel the water to a point so it runs off — that way the weight of the water doesn't damage the plant.
2) Tall trees have big roots called buttress roots to support their trunks.
3) Climbing plants, such as lianas, use the tree trunks to climb up to the light.
4) Many trees are evergreen — they keep their leaves all year round to take advantage of the continual growing season.

Rainforests — hot, wet and full of creepy bugs, eugh...

Make sure you know the climate, soil, vegetation structure and plant adaptations for the rainforest ecosystem. Cover the page and scribble down what you know to check. And give yourself bonus marks for getting the tree layer diagram right.

World Ecosystems

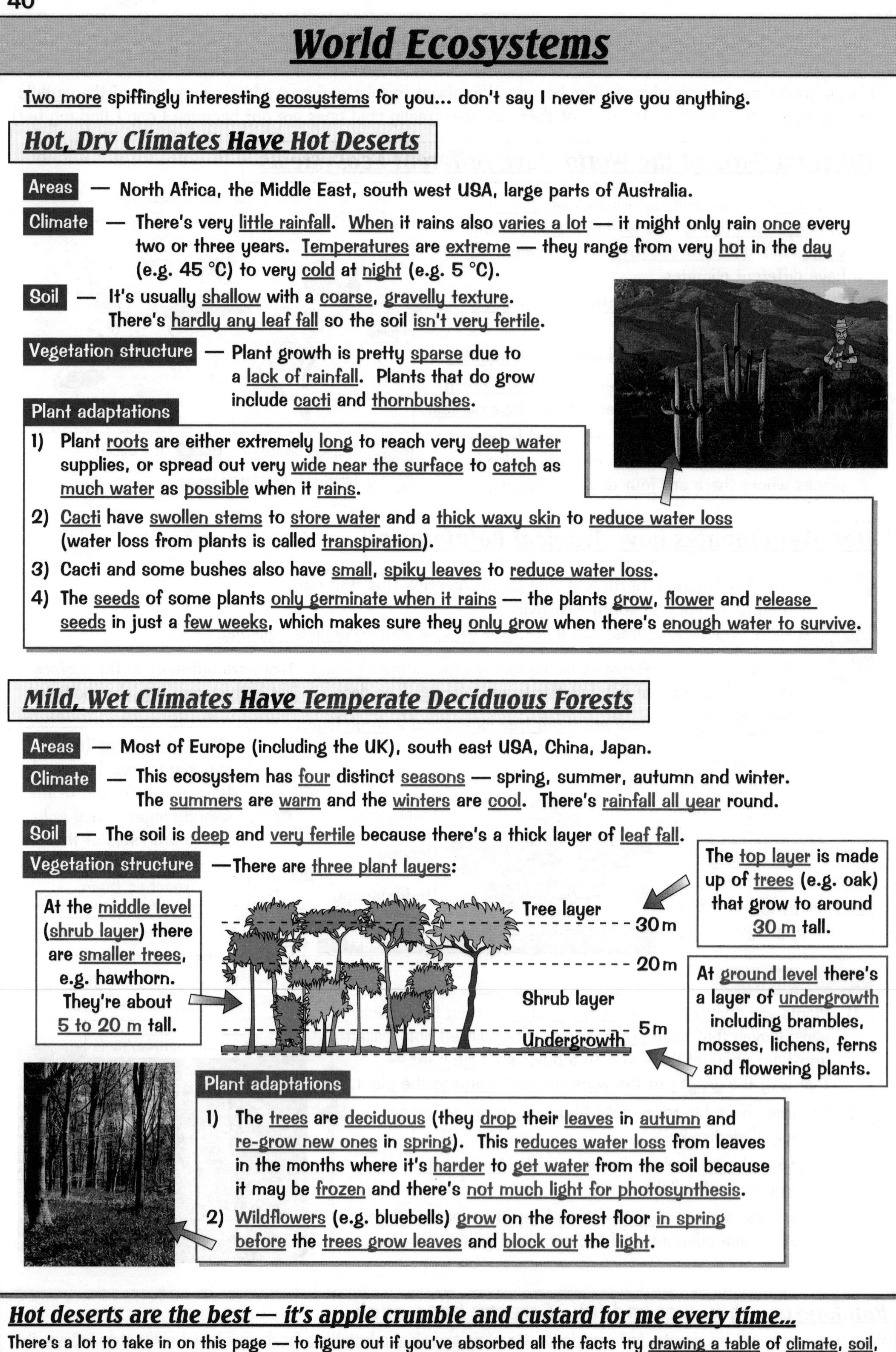

Two more spiffingly interesting ecosystems for you... don't say I never give you anything.

Hot, Dry Climates Have Hot Deserts

Areas — North Africa, the Middle East, south west USA, large parts of Australia.

Climate — There's very little rainfall. When it rains also varies a lot — it might only rain once every two or three years. Temperatures are extreme — they range from very hot in the day (e.g. 45 °C) to very cold at night (e.g. 5 °C).

Soil — It's usually shallow with a coarse, gravelly texture. There's hardly any leaf fall so the soil isn't very fertile.

Vegetation structure — Plant growth is pretty sparse due to a lack of rainfall. Plants that do grow include cacti and thornbushes.

Plant adaptations

1) Plant roots are either extremely long to reach very deep water supplies, or spread out very wide near the surface to catch as much water as possible when it rains.
2) Cacti have swollen stems to store water and a thick waxy skin to reduce water loss (water loss from plants is called transpiration).
3) Cacti and some bushes also have small, spiky leaves to reduce water loss.
4) The seeds of some plants only germinate when it rains — the plants grow, flower and release seeds in just a few weeks, which makes sure they only grow when there's enough water to survive.

Mild, Wet Climates Have Temperate Deciduous Forests

Areas — Most of Europe (including the UK), south east USA, China, Japan.

Climate — This ecosystem has four distinct seasons — spring, summer, autumn and winter. The summers are warm and the winters are cool. There's rainfall all year round.

Soil — The soil is deep and very fertile because there's a thick layer of leaf fall.

Vegetation structure —There are three plant layers:

Plant adaptations

1) The trees are deciduous (they drop their leaves in autumn and re-grow new ones in spring). This reduces water loss from leaves in the months where it's harder to get water from the soil because it may be frozen and there's not much light for photosynthesis.
2) Wildflowers (e.g. bluebells) grow on the forest floor in spring before the trees grow leaves and block out the light.

Hot deserts are the best — it's apple crumble and custard for me every time...

There's a lot to take in on this page — to figure out if you've absorbed all the facts try drawing a table of climate, soil, vegetation structure and plant adaptations for the two ecosystems. If there are any blank boxes read the page again.

Temperate Deciduous Forest — Case Study

Batten down the hatches, get your supplies ready and prepare for some details about deciduous forests...

Deciduous Forests are Used for Many Things

Forests don't just look pretty — they can be used for loads of things. For example:

1) Timber — trees are cut down and the wood is sold to make money.
2) Timber products — the wood can be processed to make products such as fencing and furniture.
3) Recreation — forests are used for walking, cycling and other outdoor activities.

If a forest is going to be used in the long-term, it has to be managed in a way that's sustainable, i.e. in a way that allows people today to get the things they need, but without stopping people in the future from getting what they need.

Forests can be Carefully Managed to Conserve Them For The Future

Here are some examples of sustainable management strategies:

1) Controlled felling — instead of clearing all the trees in an area, only some trees are cut down, e.g. trees over a certain age or just one species. This is less damaging to the forest than felling all the trees in an area because the overall forest structure is kept. This means the forest will be able to regenerate so it can be used in the future.

   Controlled felling is also called selective logging.

2) Replanting — where trees are felled, they're replaced by planting new trees. This makes sure that overall the amount of forest is not reduced and people can keep using it in the future.
3) Planning for recreational use — lots of visitors can damage a forest, for example by causing erosion, dropping litter and disturbing wildlife. Good management can reduce damage so use can continue in the future, e.g. by paving footpaths to reduce erosion and providing plenty of bins to reduce litter.

Case Study — The New Forest in Hampshire

1) The New Forest is a National Park that covers 375 km^2.
2) It's used for timber, timber products, farming and recreation.
   - The New Forest produces around 50 000 tonnes of timber a year.
   - Local mills make fencing products out of the timber from the New Forest.
   - Around 20 million visitors come to the forest per year. Recreational activities available include walking, cycling (there are over 100 miles of cycle tracks), wildlife watching (visitors particularly come to see the New Forest ponies, which roam wild), horse riding, fishing, golf, watersports and special events such as the New Forest and Hampshire County Show.
3) The forest is managed to make sure the way it's used is sustainable:
   - Areas cleared of trees are either replanted or restored to other habitats like heathland.
   - Walkers and cyclists are encouraged to stick to the footpaths and cycle paths to limit damage to surrounding habitats. Also dogs aren't allowed near wildlife breeding sites at certain times of year. These measures help to conserve wildlife so it's still there for future generations.
   - Recreational users are encouraged to act responsibly (e.g. close gates, take litter home) by information at the National Park Forest Centre and local information points.

The New Forest — a bit like the old one but bigger, faster and shinier...

Eeee, the New Forest sounds nice — a magical forest with wild ponies and pigs roaming around. Try to keep that image in your head when you're scribbling out all the details about it in the exam. Which reminds me... get learnin' the details.

Tropical Rainforest — Deforestation

Removal of trees from forests is called deforestation. It's happening on a huge scale in many tropical rainforests. Deforestation has many impacts — some good, some bad and some downright ugly.

There are Five Main Causes of Deforestation

Farming — forest is cleared to set up small subsistence farms or large commercial cattle ranches. Often the "slash and burn" technique is used to clear the forest — vegetation is cut down and left to dry then burnt.

Commercial logging — trees are felled to make money.

Population pressure — as the population in the area increases, trees are cleared to make land for new settlements.

Mineral extraction — minerals (e.g. gold and iron ore) are mined and sold to make money. Trees are cut down to expose ground and to clear access routes.

Road building — more settlements and industry (e.g. logging and mining) lead to more roads being built. Trees along the path of the road have to be cleared to build them.

Deforestation has Environmental, Social, Political and Economic Impacts

ENVIRONMENTAL

1) Fewer trees means fewer habitats and food sources for animals and birds. This reduces biodiversity as organisms either have to move or become extinct.
2) With no trees to hold the soil together, heavy rain washes away the soil (soil erosion).
3) If a lot of soil from deforested areas is washed into rivers it can kill fish, make the water undrinkable and cause flooding (as the riverbed is raised so it can't hold as much water).
4) Without a tree canopy to intercept (catch) rainfall and tree roots to absorb it, more water reaches the soil. This increases the risk of flooding and reduces soil fertility as nutrients in the soil are washed down into the earth, out of reach of plants.
5) Without trees there's no leaf fall — so no nutrient supply to the soil, which makes it less fertile.
6) Trees remove CO_2 from the atmosphere when they photosynthesise, so without them less CO_2 is removed. Also, burning vegetation to clear forest produces CO_2. So deforestation means more CO_2 in the atmosphere, which adds to global warming.
7) Without trees, water isn't removed from the soil and evaporated into the atmosphere. So fewer clouds form and rainfall in the area is reduced. Reduced rainfall reduces plant growth.

SOCIAL

1) The quality of life for some local people improves as there are more jobs.
2) The livelihoods of some local people are destroyed — deforestation can cause the loss of the animals and plants that they rely on to make a living.
3) Some native tribes have been forced to move when trees on their land have been cleared.
4) There can be conflict between native people, landowners, mining companies and logging companies over use of land.

ECONOMIC

1) Logging, farming and mining create jobs.
2) A lot of money is made from selling timber, mining and commercial farming.

POLITICAL

There's pressure from foreign governments to stop deforestation.

The deforestation revision page — it's a cut above the rest...

A bit of a serious page, this one. I would try to brighten it up by doing a dance or something but you'd probably just point and laugh. So I'll stay sitting down and recommend you learn this page — deforestation is a fave with examiners.

Tropical Rainforest — Sustainable Management

It's not all doom and gloom for rainforests. In fact, this page is dedicated to the ways to manage them.

Tropical Rainforests can be Sustainably Managed

Rainforests can be managed in a way that's sustainable, i.e. in a way that allows people today to get the things they need, but without stopping people in the future from getting what they need. Here are some of the ways it can be done:

1 Selective Logging

1) Only some trees (e.g. just the oldest ones) are felled — most trees are left standing.
2) This is less damaging to the forest than felling all the trees in an area. If only a few trees are taken from each area the overall forest structure is kept — the canopy's still there and the soil isn't exposed. This means the forest will be able to regenerate so it can be used in the future.
3) The least damaging forms are 'horse logging' and 'helicopter logging' — dragging felled trees out of the forest using horses or removing them with helicopters instead of huge trucks.

EXAMPLE: Helicopter logging is used in the Malaysian state of Sarawak.

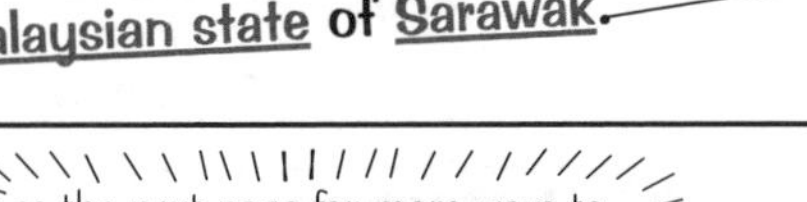

See the next page for more ways to manage rainforests sustainably.

3 Reducing Demand for Hardwood

1) Hardwood is a general term for wood from certain tree species, e.g. mahogany and teak. The wood tends to be fairly dense and hard — it's used to make things like furniture.
2) There's a high demand for hardwood from consumers in richer countries.
3) This means that some tropical hardwood trees are becoming rarer as people are chopping them down and selling them.
4) Some richer countries are trying to reduce demand so fewer of these tree species are cut down, which means they'll exist for future generations to use.
5) Strategies to reduce demand include heavily taxing imported hardwood or banning its sale.
6) Some countries with tropical rainforests also ban logging of hardwood species.

2 Replanting

1) This is when new trees are planted to replace the ones that are cut down.
2) This means there will be trees for people to use in the future.
3) It's important that the same types of tree are planted that were cut down, so that the variety of trees is kept for the future.
4) In some countries there are environmental laws to make logging companies replant trees when they clear an area.

4 Education

1) Some local people don't know what the environmental impacts of deforestation are. Local people try to make money in the short-term (e.g. by illegal logging) to overcome their own poverty.
2) Educating these people about the impacts of deforestation and ways to reduce the impacts decreases their effect on the environment.
3) Also, educating them about alternative ways to make money that don't damage the environment, e.g. ecotourism (see the next page), reduces their impact.
4) Both of these things mean that the rainforest is conserved and so will be there for future generations to use.
5) Education of the international community about the impacts of deforestation will reduce demand for products that lead to deforestation, e.g. hardwood furniture. It will also put pressure on governments to reduce deforestation.

Selective logging — a bit more precise than eeny, meeny, miny, mo...

Told you it wasn't all doom and gloom. You need to really get your head around what sustainable management is and how it makes sure that there are loads of trees, orangutans, tarantulas and other creepy bugs for future generations.

Tropical Rainforest — Sustainable Management

More rainforest sustainable management strategies for you to sink your teeth into...

Tropical Rainforests can be Sustainably Managed

1) Ecotourism

1) Ecotourism is tourism that doesn't harm the environment and benefits the local people.
2) Ecotourism provides a source of income for local people, e.g. they act as guides, provide accommodation and transport.
3) This means the local people don't have to log or farm to make money. So fewer trees are cut down, which means there are more trees for the future.
4) Ecotourism is usually a small-scale activity, with only small numbers of visitors going to an area at a time. This helps to keep the environmental impact of tourism low.
5) Ecotourism should cause as little harm to the environment as possible. For example, by making sure waste and litter are disposed of properly to prevent land and water contamination.
6) Ecotourism helps the sustainable development of an area because it improves the quality of life for local people without stopping people in the future getting what they need (because it doesn't damage the environment or deplete resources).

EXAMPLE: Tataquara Lodge is a tourist lodge in the Brazilian rainforest. The lodge has 15 rooms and offers activities like fishing, canoeing, wildlife viewing and forest walks. Waste is disposed of responsibly and it runs lights using solar power.

2) Reducing Debt

1) A lot of tropical rainforests are in poorer countries, e.g. Nigeria, Belize and Burma.
2) Poorer countries often borrow money from richer countries or organisations (e.g. the World Bank) to fund development schemes or cope with emergencies like floods.
3) This money has to be paid back (sometimes with interest).
4) These countries often allow logging, farming and mining in rainforests to make money to pay back the debt.
5) So reducing debt would mean countries wouldn't have to do this and the rainforests could be conserved for the future.
6) Debt can be cancelled by countries or organisations, but there's no guarantee the money will be spent on conservation.
7) Conservation swaps (debt-for-nature swaps) guarantee the money is spent on conservation — part of a country's debt is paid off by someone else in exchange for investment in conservation.

EXAMPLE: In 1987 a conservation group paid off some of Bolivia's debt in exchange for creating a rainforest reserve.

See the previous page for a definition of sustainable management.

3) Protection

1) Environmental laws can be used to protect rainforests. For example:
   - Laws that ban the use of wood from forests that are managed non-sustainably.
   - Laws that ban illegal logging.
   - Laws that ban logging of some tree species, e.g. mahogany.
2) Many countries have set up national parks and nature reserves within rainforests. In these areas damaging activities, e.g. logging, are restricted. However, a lack of funds can make it difficult to police the restrictions.

Rainforest protection — you'll need an umbrella and some wellies...

Don't forget — the basic idea is that anything that allows people today to get what they need whilst stopping the rainforest being damaged or its resources being depleted is sustainable management.

Tropical Rainforest — Case Study

The Amazon is the largest rainforest on Earth, but it's shrinking fast due to deforestation.

Deforestation is a Problem in the Amazon

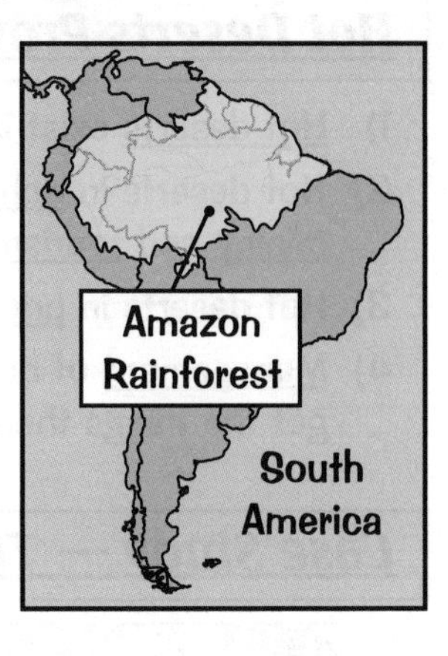

The Amazon covers an area of around 8 million km^2, including parts of Brazil, Peru, Colombia, Venezuela, Ecuador, Bolivia, Guyana, Suriname and French Guiana. However, since 1970 over 600 000 km^2 has been destroyed by deforestation. There are lots of causes — for example, between 2000 and 2005:

1) 60% was caused by cattle ranching.
2) 33% was caused by small-scale subsistence farming.
3) 3% was caused by logging.
4) 3% was caused by mining, urbanisation, road construction, dams and fires.
5) 1% was caused by large-scale commercial farming (other than cattle ranching).

Deforestation in the Amazon has Many Impacts

Environmental	• Habitat destruction and loss of biodiversity, e.g. the number of endangered species in Brazil increased from 218 in 1989 to 628 in 2008. • The Amazon stores around 100 billion tonnes of carbon — deforestation will release some of this as carbon dioxide, which causes global warming.
Social	• Local ways of life have been affected, e.g. some Brazilian rubber tappers have lost their livelihoods as rubber trees have been cut down. • Native tribes have been forced to move, e.g. some of the Guarani tribe in Brazil have moved because their land was taken for cattle ranching and sugar plantations. • There's conflict between large landowners, subsistence farmers and native people, e.g. in 2009 there were riots in Peru over rainforest destruction and hundreds of native Indians were killed or injured.
Economic	• Farming makes a lot of money for countries in the rainforest, e.g. in 2008, Brazil made $6.9 billion from trading cattle. • The mining industry creates jobs for loads of people, e.g. the Buenaventura Mining Company in Peru employs over 3100 people.

Several Sustainable Management Strategies are being Used

1) Some deforested areas are being replanted with new trees, e.g. Peru plans to replant more than 100 000 km^2 of forest before 2018.
2) Some countries are trying to reduce the number of hardwood trees felled, e.g. Brazil banned mahogany logging in 2001 and seizes timber from illegal logging companies.
3) Ecotourism is becoming more popular, e.g. the Madre de Dios region in Peru has around 70 lodges for ecotourists — 60 000 people visited the region in 2007.
4) Most countries have environmental laws to help protect the rainforest, e.g. the Brazilian Forest Code says that landowners have to keep 50-80% of their land as forest.
5) Some countries have national parks, e.g. the Central Amazon Conservation Complex in Brazil is the largest protected area in the rainforest, covering around 25 000 km^2. It's a World Heritage Site that's home to loads of ecosystems and animals like black caimans and river dolphins.
6) Reducing debt has helped some countries conserve their rainforest, e.g. in 2008 the USA reduced Peru's debt by $25 million in exchange for conserving its rainforest.

Tarzan wouldn't have much fun in the Amazon...

At this point I'm supposed to write something helpful, but it turns out I'm a bit short of ideas. I've even tried making a cup of tea to help the ideas flow. I'll be prepared with a nugget of wisdom by the time you finish page 46 though.

Hot Deserts — Case Study

Hot deserts aren't just places for riding camels — believe it or not, they're used for loads of stuff.

Hot Deserts Provide Economic Opportunities

See the next page for another hot desert case study.

1) Hot deserts exist in rich and poor areas of the world.
2) Hot deserts in rich areas are usually used for things like commercial farming, mining and tourism. Lots of people also retire there (retirement migration).
3) Hot deserts in poor areas are usually used for hunting and gathering and farming.
4) Management of both rich and poor deserts needs to be sustainable — i.e. to allow people today to get the things they need, but without stopping people in the future from getting what they need.

Case Study — The Kalahari Desert is a Relatively Poor Region

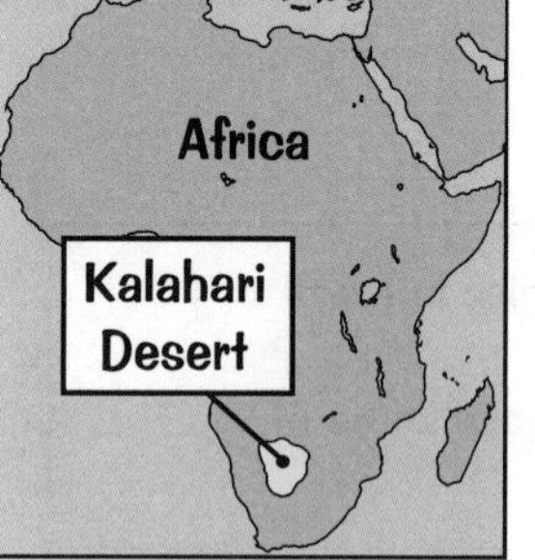

1) The Kalahari Desert has an area of 260 000 km^2. It covers most of Botswana and parts of Zimbabwe, Namibia and South Africa.
2) It gets little rain (about 200 mm per year). The only permanent river in the area is the Boteti River. However, temporary streams and rivers form after rain. The low rainfall in the area means that droughts are a problem.
3) The Kalahari is very sparsely populated, but there are native people that live there, e.g. the San Bushmen and the Tswana. Some native people still hunt wild game (e.g. antelope) with bows and arrows and gather plants for food.
4) Farming cattle, goats and sheep is a big industry in the Kalahari, e.g. in 1998 there were 2.3 million cattle in Botswana. Some grazing land is irrigated using groundwater from boreholes.
5) There's lots of mining in the area — there are coal, diamond, gold, copper, nickel and uranium mines, e.g. the Opara Diamond Mine in Botswana.
6) Some uses of the Kalahari have negative impacts:

   1) Overgrazing of land has caused soil erosion, and irrigation has depleted groundwater supplies.
   2) Fences put up by farmers have blocked migration routes of wild animals, e.g. wildebeest. The animals can't move to where the grazing is best so some die from starvation.
   3) Mining and farming have led to native people being forced off their land.
   4) Mining uses a lot of water from boreholes. This is depleting groundwater supplies.

7) Here are a few of the management strategies being carried out in the Kalahari:

   1) Some places are trying to conserve water. E.g. in Windhoek in Namibia people are charged for the volume of water they use. This encourages them to use less. This is more sustainable because water supplies aren't depleted as much and so more will be there in the future.
   2) Water supply all over the Kalahari is being increased by building dams and drilling more boreholes. This allows more farming and reduces the effects of drought, but isn't sustainable because it depletes groundwater supplies even more.
   3) Several game reserves have been created to provide areas for the native people to live and to protect wildlife. For example, the Central Kalahari Game Reserve in Botswana was set up in 1961 as a refuge for the San Bushmen. This is sustainable because it conserves the way of life of the native people and conserves the wildlife for future generations.
   4) Some agricultural fences have been removed to allow animals to migrate. This is sustainable because fewer wild animals die so they will still be around in the future.

The Kalahari Desert — isn't that full of squid...

I know, I know, another case study. They're absolutely everywhere I'm afraid. That doesn't mean you should skip over learning them though. Cover the page and see what's stuck in your head. Better to find out what you know now...

Hot Deserts — Case Study

Keep going my desert friend, just one more case study to go before the oasis.

Case Study — The Mojave Desert is a Relatively Rich Region

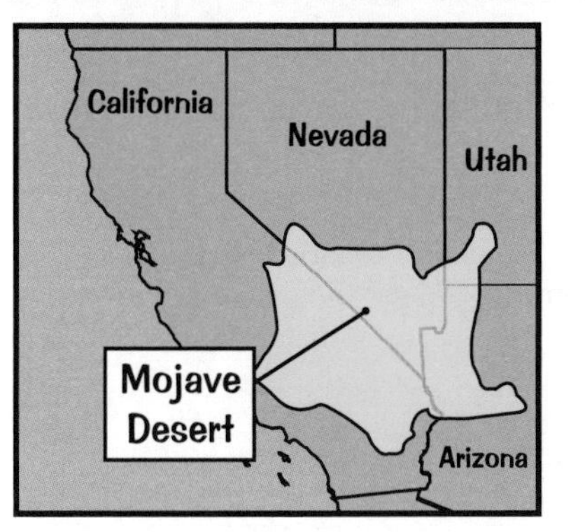

1) The Mojave Desert in the USA covers over 57 000 km^2 and includes parts of California, Nevada, Utah and Arizona.
2) The region gets less than 250 mm of rain per year.
3) There's commercial farming in the area. For example, there have been cattle ranches in the region for over 100 years.
4) The area is sparsely populated but the population is increasing, e.g. Las Vegas in Nevada is the USA's fastest growing city. The area is popular with people retiring due to its year-round good weather, e.g. 80% of the people in Sun City in Arizona are over 65.

5) Water for farming and for people comes from groundwater, the Mojave River and the Colorado River.
6) The region has many tourist destinations including Las Vegas, Death Valley and the Grand Canyon. The Death Valley National Park gets around 1 million visitors per year. Tourists are attracted by the wildlife and geology, and activities like camping, hiking, horse riding and off-road driving.
7) In the past, gold, silver, copper, lead and salts were mined, although most mines have now closed. There are a few borax mines still working in California.
8) Some uses of the Mojave have negative impacts:

   1) Rapid population growth (including retirement migrants) has depleted water resources.
   2) Farming uses a lot of water, and it can also cause soil erosion.
   3) Tourists deplete water resources, drop litter, damage plants and cause soil erosion (e.g. by using off-road vehicles).

9) Here are a few of the management strategies being carried out in the Mojave:

   1) There are water conservation schemes in the area, e.g. the Mojave Water Agency gives people vouchers to buy water efficient toilets and washing machines. They also pay people to remove grass lawns (which need a lot of water) and replace them with plants that don't use as much water. These things are more sustainable because they don't deplete water supplies as much, so there's more for future generations.
   2) The Mojave Desert has four National Parks (Death Valley, Joshua Tree, Zion and the Grand Canyon). Native species are protected and there are strict rules on land use, e.g. there are strict rules on mining to reduce environmental damage. This is sustainable because it conserves the area, so future generations can use it.
   3) There are designated roads for off-road vehicles, and sensitive areas are fenced off so they can't get in. This is sustainable because it helps conserve the plant life for future generations.
   4) Some hotels in Las Vegas are trying to conserve water, e.g. the MGM Mirage® Hotels use drip-irrigation to water lawns. This is more sustainable as it doesn't use as much water as other irrigation methods, so conserves more water for the future.

Death Valley — sounds like a laugh a minute...

Mojave (said 'mo-har-ve') means 'the meadows', which is a bit weird as not much grows there. It's also a bit weird that something like water can run out, but hot deserts don't get a lot of it so if people use it up, it might not get replaced.

Revision Summary for Section 4

So now you know absolutely everything there is to know about ecosystems. Or at least you know everything you need to for the exam. But before you go rushing off to celebrate with a slice of pumpkin pie and an Irish jig, best make sure you actually do know it. Now's as good a time as any, so give these questions a go.

1) Define the term ecosystem.
2) What is a consumer?
3) What is a food web?
4) Describe how nutrients are transferred to the soil in an ecosystem.
5) Describe the global distribution of tropical rainforests.
6) How many layers of vegetation does a tropical rainforest have?
7) What is an emergent tree?
8) Give three ways rainforest plants are adapted to their environment.
9) Describe the soil in a hot desert.
10) Give two ways plants are adapted to the hot desert environment.
11) Describe the climate of a temperate deciduous forest.
12) How tall is the top tree layer in a temperate deciduous forest?
13) What is controlled felling?
14) a) Give an example of a temperate deciduous forest.
    b) Describe how the forest is used for recreation.
    c) Describe how the forest is managed to make sure the way it's used is sustainable.
15) What are the five main causes of rainforest deforestation?
16) Give three environmental impacts of rainforest deforestation.
17) Give two social impacts of rainforest deforestation.
18) What is selective logging?
19) What is replanting?
20) How does reducing demand for hardwood help to conserve rainforests?
21) How can education be used to reduce rainforest deforestation?
22) What is ecotourism?
23) Give one way a country can reduce its debt in order to reduce deforestation.
24) Give two ways a country can protect its rainforest.
25) a) Give an example of a tropical rainforest.
    b) Describe one social, one economic and one environmental impact of deforestation in that rainforest.
    c) Give three ways the rainforest is being sustainably managed.
26) a) Give an example of a hot desert in a rich part of the world and a hot desert in a poorer part of the world.
    b) Compare the way the two desert areas are used.
    c) How is the desert in the rich area sustainably managed?

The Hydrological Cycle

Since this is the first page of a shiny new section I'm going to be nice and treat you to something special — the hydrological cycle (a.k.a. the water cycle in non-geography lingo). Knock yourself out...

The Hydrological Cycle Shows How Water Moves Around

1) The hydrological cycle has different parts — the sea, the land and the atmosphere.
2) Water flows between the different parts in various ways, and is also stored on the land (see below).
3) The hydrological cycle is a closed system. This means there are no inputs (water going in) or outputs (water going out) — the water just flows round and round the cycle:

(1) Water evaporates from the sea and the land — evaporation is when water is heated by the sun and turns into water vapour. Transpiration is the evaporation of water from plants. Evapotranspiration is both evaporation and transpiration happening together.

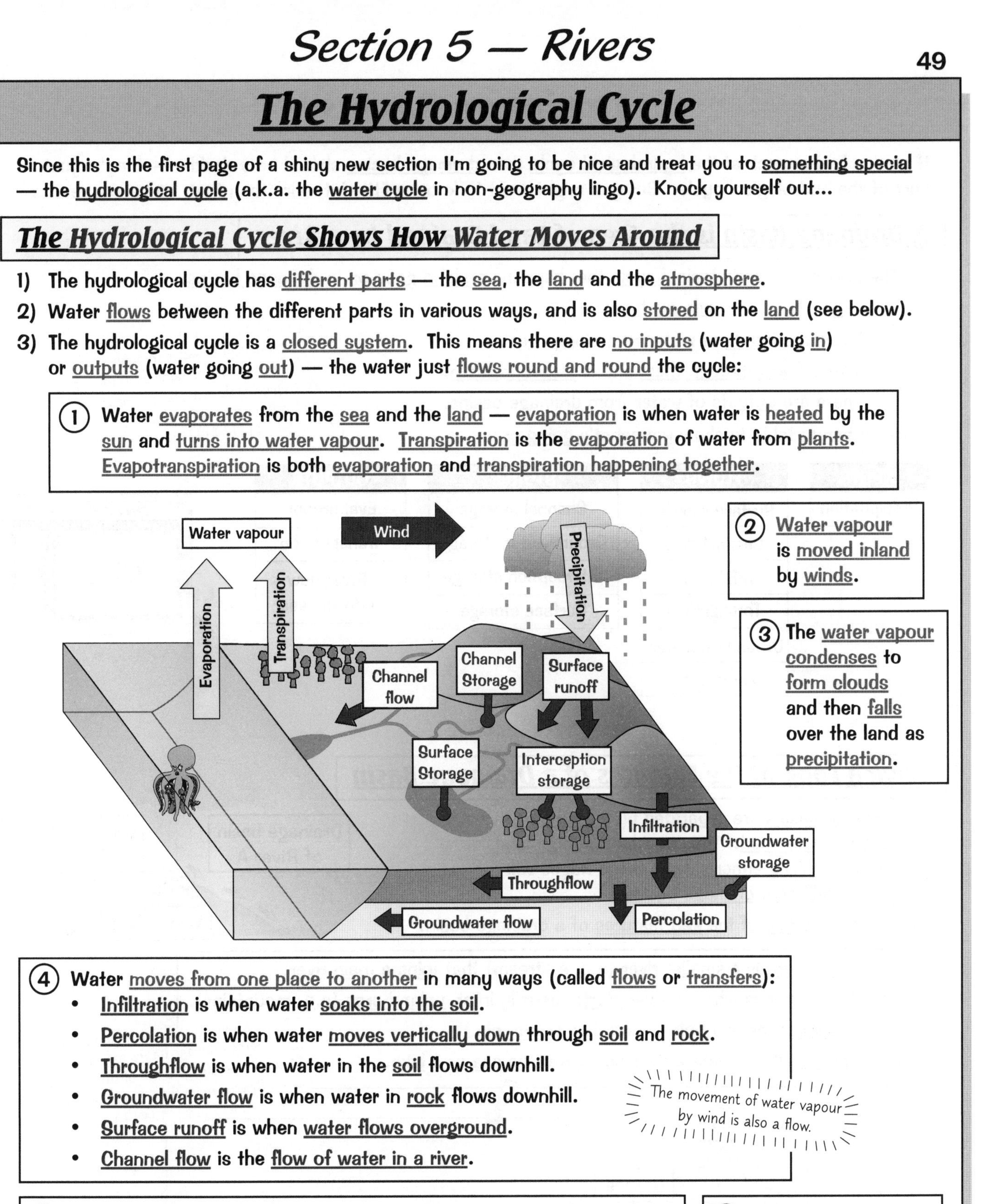

(2) Water vapour is moved inland by winds.

(3) The water vapour condenses to form clouds and then falls over the land as precipitation.

(4) Water moves from one place to another in many ways (called flows or transfers):

- Infiltration is when water soaks into the soil.
- Percolation is when water moves vertically down through soil and rock.
- Throughflow is when water in the soil flows downhill.
- Groundwater flow is when water in rock flows downhill.
- Surface runoff is when water flows overground.
- Channel flow is the flow of water in a river.

The movement of water vapour by wind is also a flow.

(5) Water can also be held on the land in stores:

- Channel storage is when water is held in a river.
- Groundwater storage is when water is stored underground in soil and rock. A rock that stores water is called an aquifer, e.g. chalk.
- Interception storage is when water lands on things like plant leaves and doesn't hit the ground.
- Surface storage is when water is held in things like lakes, reservoirs and puddles.

(6) The water eventually ends up in the sea, where it evaporates and goes round the cycle again...

Percolation and infiltration — sounds like a recipe for a great cuppa...

Crikey, there are a lot of geography terms to get your head round on this page — make sure you understand what each one means. To check, shut the book and scribble out as many as you can remember (no sneaky peeking either).

Drainage Basins

If you've had a crack at the hydrological cycle then drainage basins should be a doddle — they're just a part of the hydrological cycle. But you might want to know a little bit more than that, so have a read...

A Drainage Basin is the Area of Land Drained by a River

1) The part of the hydrological cycle that happens on land goes on in drainage basins.
2) Drainage basins are open systems:
   - There are inputs of water to drainage basins.
   - Water flows through them and is stored in them.
   - There are outputs of water from drainage basins.

The movement of water through a drainage basin is the same as in the hydrological cycle (see previous page), but without the sea and wind bit.

Here's a handy table to show you what's going on:

INPUTS	FLOWS	STORES	OUTPUTS
Precipitation	Surface runoff	Channel storage	Evaporation
	Channel flow	Groundwater storage	Transpiration
	Infiltration	Interception storage	River flow into the sea
	Throughflow	Surface storage	
	Groundwater flow		
	Percolation		

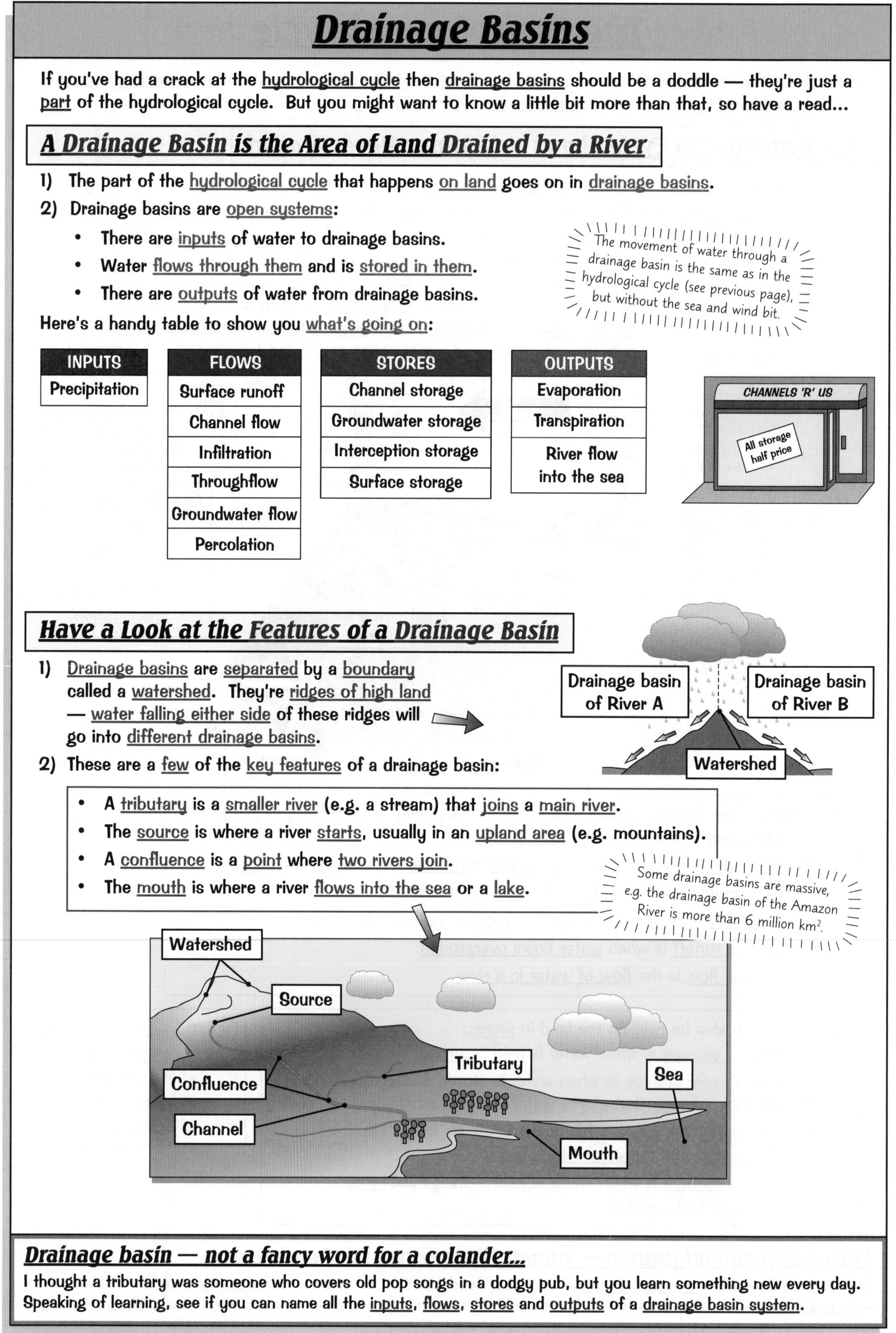

Have a Look at the Features of a Drainage Basin

1) Drainage basins are separated by a boundary called a watershed. They're ridges of high land — water falling either side of these ridges will go into different drainage basins.
2) These are a few of the key features of a drainage basin:
   - A tributary is a smaller river (e.g. a stream) that joins a main river.
   - The source is where a river starts, usually in an upland area (e.g. mountains).
   - A confluence is a point where two rivers join.
   - The mouth is where a river flows into the sea or a lake.

Some drainage basins are massive, e.g. the drainage basin of the Amazon River is more than 6 million km^2.

Drainage basin — not a fancy word for a colander...

I thought a tributary was someone who covers old pop songs in a dodgy pub, but you learn something new every day. Speaking of learning, see if you can name all the inputs, flows, stores and outputs of a drainage basin system.

Weathering and the River Valley

You won't find much chocolate in a drainage basin, but you will find rocks, river valleys and ramblers...

Rocks in a Drainage Basin are Broken Down by Weathering

Weathering happens in drainage basins — it's the breakdown of rocks where they are (the material created doesn't get taken away like with erosion). There are three main types of weathering:

1) Mechanical weathering is the breakdown of rock without changing its chemical composition. Freeze-thaw weathering is a type of mechanical weathering that happens in drainage basins:

   1) It happens when the temperature alternates above and below 0 °C (the freezing point of water).
   2) Water gets into rock that has cracks, e.g. granite.
   3) When the water freezes it expands, which puts pressure on the rock.
   4) When the water thaws it contracts, which releases the pressure on the rock.
   5) Repeated freezing and thawing widens the cracks and causes the rock to break up.

2) Chemical weathering is the breakdown of rock by changing its chemical composition. Carbonation weathering is a type of chemical weathering that happens in warm and wet conditions:

   1) Rainwater has carbon dioxide dissolved in it, which makes it a weak carbonic acid.
   2) Carbonic acid reacts with rock that contains calcium carbonate, e.g. limestone, so the rocks are dissolved by the rainwater.

3) Biological weathering is the breakdown of rocks by living things, e.g. plant roots break down rocks by growing into cracks on their surfaces and pushing them apart.

A River's Long Profile and Cross Profile Vary Over its Course

1) The path of a river as it flows downhill is called its course.
2) Rivers have an upper course (closest to the source of the river), a middle course and a lower course (closest to the mouth of the river).
3) Rivers flow in channels in valleys.
4) They erode the landscape — wear it down, then transport the material to somewhere else where it's deposited.
5) The shape of the valley and channel changes along the river depending on whether erosion or deposition is having the most impact (is the dominant process).
6) The long profile of a river shows you how the gradient (steepness) changes over the different courses.
7) The cross profile shows you what a cross-section of the river looks like.

Valley
Channel
Water

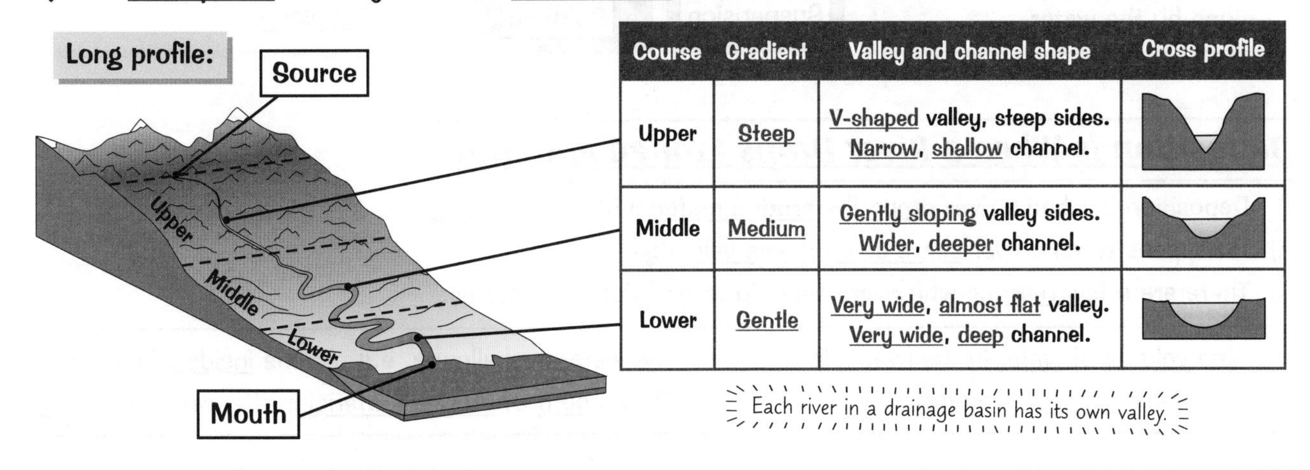

Course	Gradient	Valley and channel shape	Cross profile
Upper	Steep	V-shaped valley, steep sides. Narrow, shallow channel.	
Middle	Medium	Gently sloping valley sides. Wider, deeper channel.	
Lower	Gentle	Very wide, almost flat valley. Very wide, deep channel.	

Each river in a drainage basin has its own valley.

The river valet was rubbish at his job — all the cars got soaked...

There seems like a lot on this page but it's all pretty straightforward. Try drawing the cross profile diagrams and describing the shape of the valley and channel until you know river profiles like the back of your hand.

Erosion, Transportation and Deposition

Rivers scrape and smash rocks up, push them about, then dump them when they've had enough...

Vertical and Lateral Erosion Change the Cross Profile of a River

Erosion can be vertical or lateral — both types happen at the same time, but one is usually dominant over the other at different points along the river:

Vertical erosion

This deepens the river valley (and channel), making it V-shaped. It's dominant in the upper course of the river.

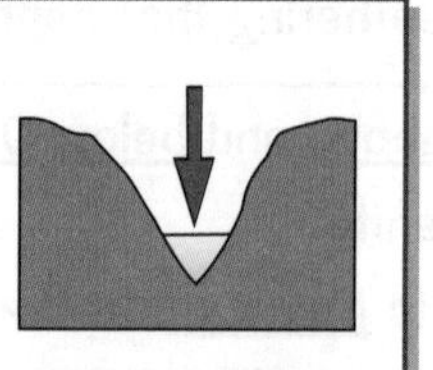

Lateral erosion

This widens the river valley (and channel). It's dominant in the middle and lower courses.

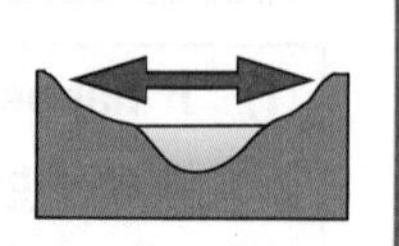

The faster a river's flowing, the more erosion happens.

There are Four Processes of Erosion

1) Hydraulic action — The force of the water breaks rock particles away from the river channel.
2) Abrasion — Eroded rocks picked up by the river scrape and rub against the channel, wearing it away. Most erosion happens by abrasion.
3) Attrition — Eroded rocks picked up by the river smash into each other and break into smaller fragments. Their edges also get rounded off as they rub together.
4) Solution — River water dissolves some types of rock, e.g. chalk and limestone.

Abrasion is sometimes called corrasion.

Transportation is the Movement of Eroded Material

The material a river has eroded is transported downstream.
There are four processes of transportation:

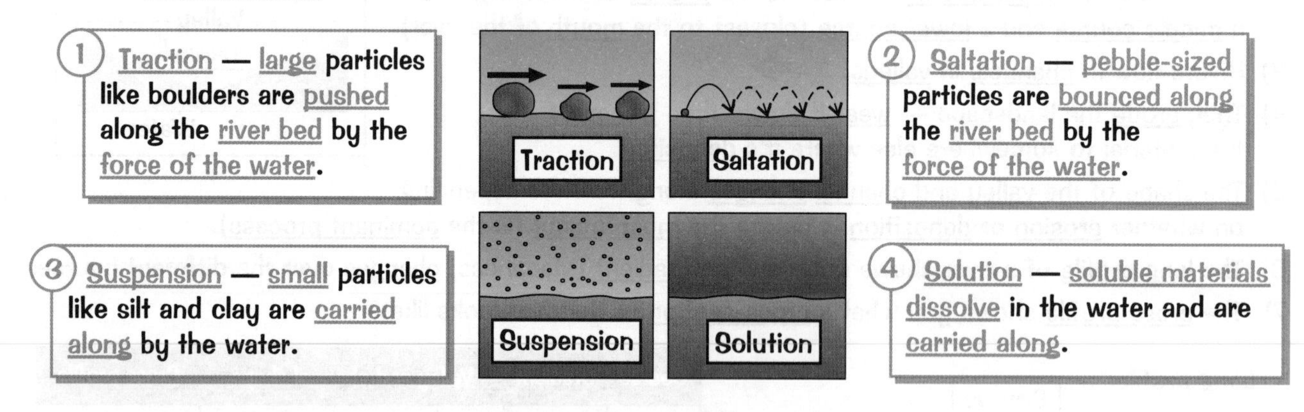

1) Traction — large particles like boulders are pushed along the river bed by the force of the water.

2) Saltation — pebble-sized particles are bounced along the river bed by the force of the water.

3) Suspension — small particles like silt and clay are carried along by the water.

4) Solution — soluble materials dissolve in the water and are carried along.

Deposition is When a River Drops Eroded Material

1) Deposition is when a river drops the eroded material it's transporting.
2) It happens when a river slows down (loses velocity).
3) There are a few reasons why rivers slow down and deposit material:

- The volume of water in the river falls.
- The river reaches its mouth.
- The water is shallower, e.g. on the inside of a bend.
- The amount of eroded material in the water increases.

In rock school there's only one punishment for naughty clay — suspension...

There are loads of amazingly similar names to remember here — try not to confuse saltation, solution and suspension. And yes, solution is both a process of erosion and transportation. Now saltate on over to the next page...

Erosional River Landforms

If you don't know anything about waterfalls then you haven't been watching enough shampoo adverts. Now's your chance to find out all about them and other landforms made by erosion.

Waterfalls and Gorges are Found in the Upper Course of a River

1) Waterfalls (e.g. High Force waterfall on the River Tees) form where a river flows over an area of hard rock followed by an area of softer rock.
2) The softer rock is eroded more than the hard rock, creating a 'step' in the river.
3) As water goes over the step it erodes more and more of the softer rock.
4) A steep drop is eventually created, which is called a waterfall.

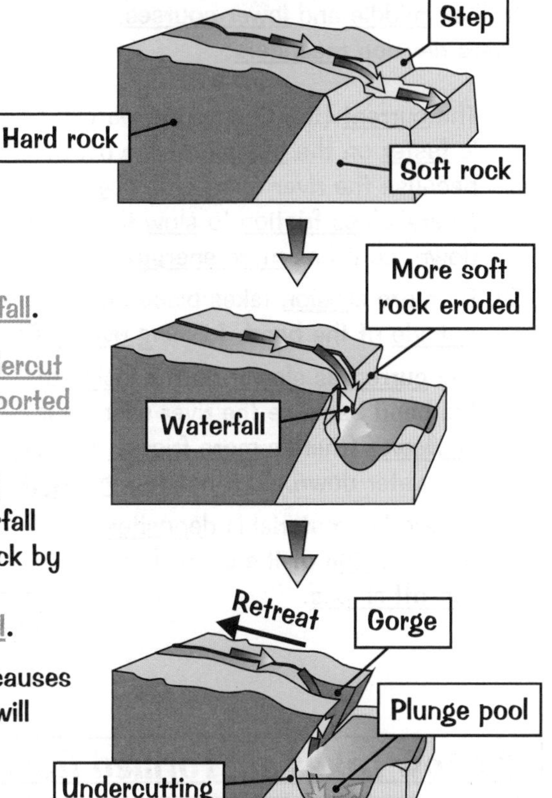

5) The hard rock is eventually undercut by erosion. It becomes unsupported and collapses.
6) The collapsed rocks are swirled around at the foot of the waterfall where they erode the softer rock by abrasion (see previous page). This creates a deep plunge pool.
7) Over time, more undercutting causes more collapses. The waterfall will retreat (move back up the channel), leaving behind a steep-sided gorge.

Interlocking Spurs are Nothing to do with Cowboys

1) In the upper course of a river most of the erosion is vertically downwards. This creates steep-sided, V-shaped valleys.
2) The rivers aren't powerful enough to erode laterally (sideways) — they have to wind around the high hillsides that stick out into their paths on either side.
3) The hillsides that interlock with each other (like a zip if you were looking from above) as the river winds around them are called interlocking spurs.

Interlocking spurs along a river in Georgia

Some river landforms are beautiful — others are gorge-ous...

Step over the hard rock and plunge into the pool — that's how I remember how waterfalls are formed. Geography examiners love river landforms (they're a bit weird like that) so it's worth knowing how they form.

Erosional and Depositional River Landforms

When a river's eroding and depositing material, meanders and ox-bow lakes can form. Australians have a different name for ox-bow lakes — billabongs. Stay tuned for more incredible facts.

Meanders are Large Bends in a River

In their middle and lower courses, rivers develop meanders:

1) The current (the flow of the water) is faster on the outside of the bend because the river channel is deeper (there's less friction to slow the water down, so it has more energy).
2) So more erosion takes place on the outside of the bend, forming river cliffs.
3) The current is slower on the inside of the bend because the river channel is shallower (there's more friction to slow the water down, so it has less energy).
4) So eroded material is deposited on the inside of the bend, forming slip-off slopes.

The Mississippi River in the USA has lots of meanders.

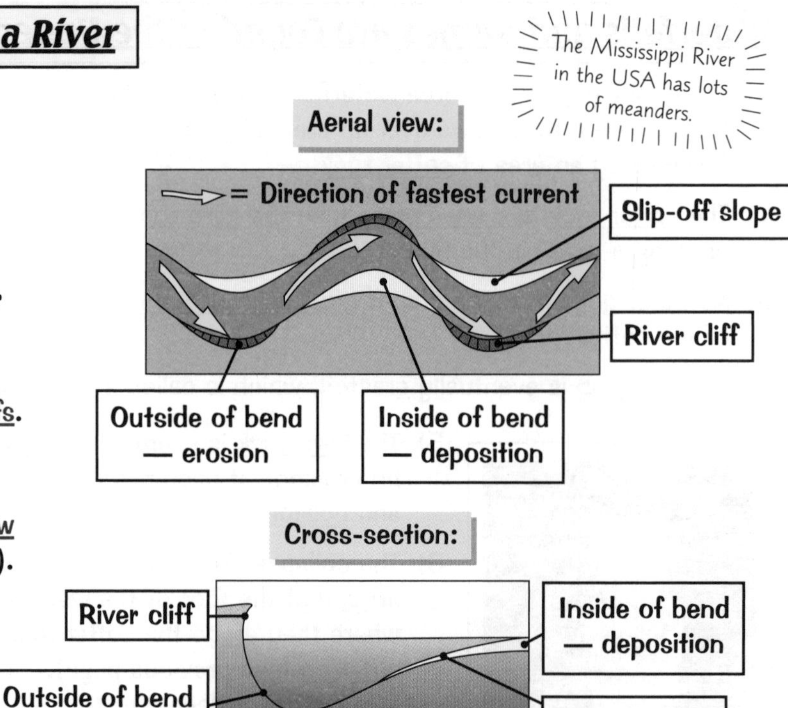

Ox-Bow Lakes are Formed from Meanders

Meanders get larger over time — they can eventually turn into an ox-bow lake:

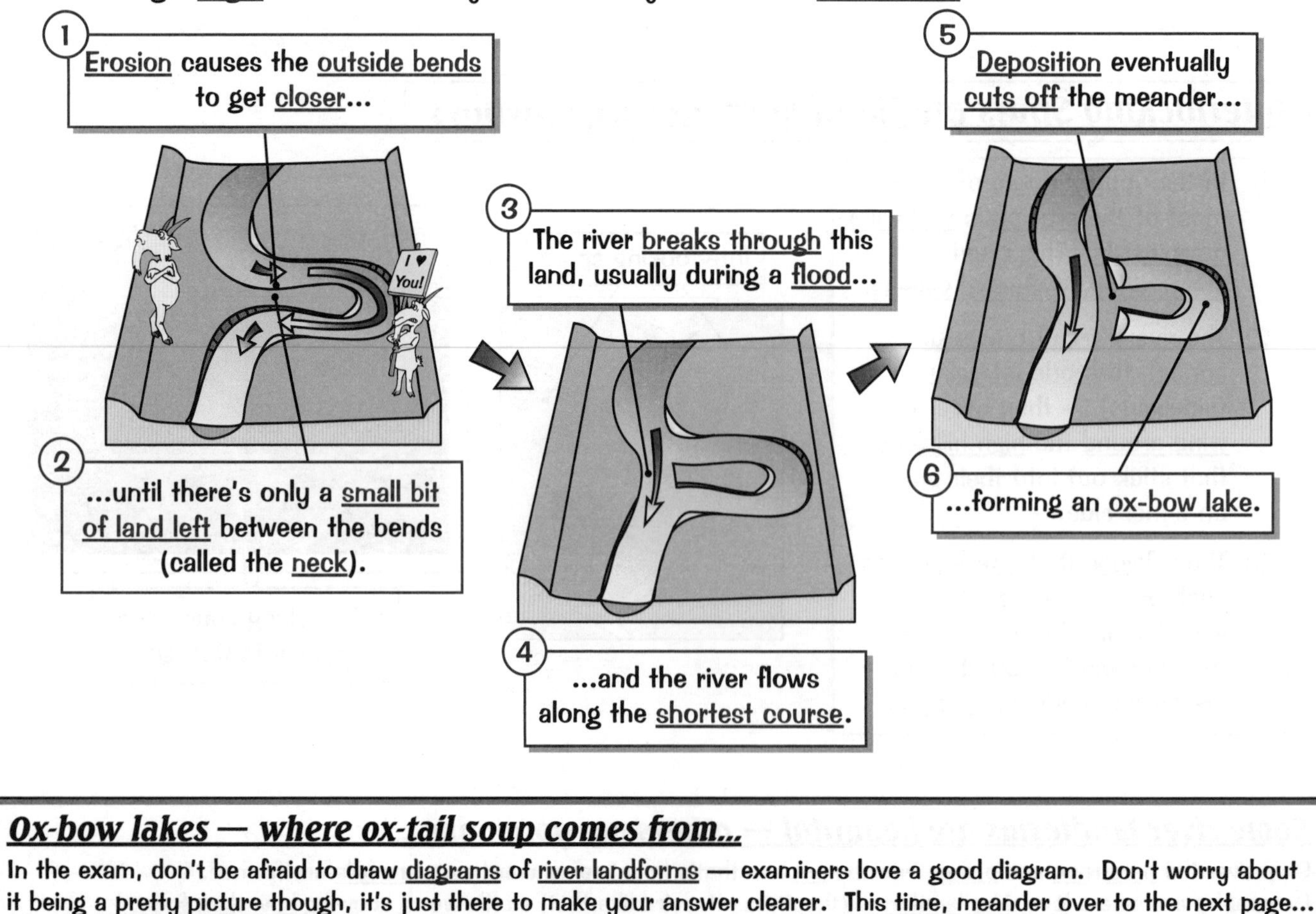

Ox-bow lakes — where ox-tail soup comes from...

In the exam, don't be afraid to draw diagrams of river landforms — examiners love a good diagram. Don't worry about it being a pretty picture though, it's just there to make your answer clearer. This time, meander over to the next page...

Depositional River Landforms

When rivers dump material they don't do it by text message — they make attractive landforms instead.

Flood Plains are Flat Areas of Land that Flood

1) The flood plain is the wide valley floor on either side of a river which occasionally gets flooded.
2) When a river floods onto the flood plain, the water slows down and deposits the eroded material that it's transporting. This builds up the flood plain (makes it higher).
3) Meanders migrate (move) across the flood plain, making it wider.
4) The deposition that happens on the slip-off slopes of meanders also builds up the flood plain.

All these landforms are found in the lower course of a river.

Levees are Natural Embankments

Levees are natural embankments (raised bits) along the edges of a river channel. During a flood, eroded material is deposited over the whole flood plain. The heaviest material is deposited closest to the river channel, because it gets dropped first when the river slows down. Over time, the deposited material builds up, creating levees along the edges of the channel, e.g. along the Yellow River in China.

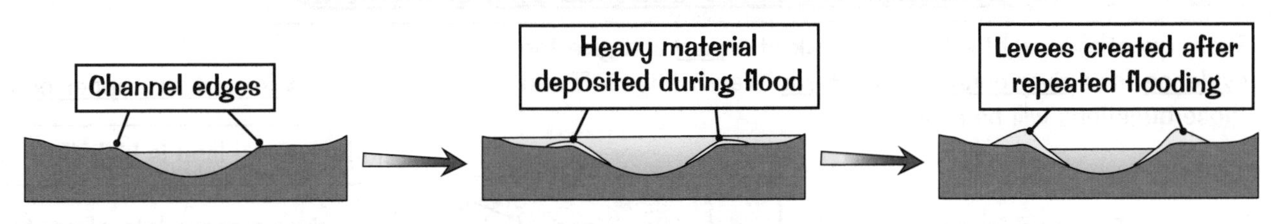

Deltas are Low-Lying Areas Where a River Meets the Sea or a Lake

1) Rivers are forced to slow down when they meet the sea or a lake. This causes them to deposit the material that they're carrying.
2) If the sea doesn't wash away the material it builds up and the channel gets blocked. This forces the channel to split up into lots of smaller rivers called distributaries.
3) Eventually the material builds up so much that low-lying areas of land called deltas are formed.
4) There are three types of delta:

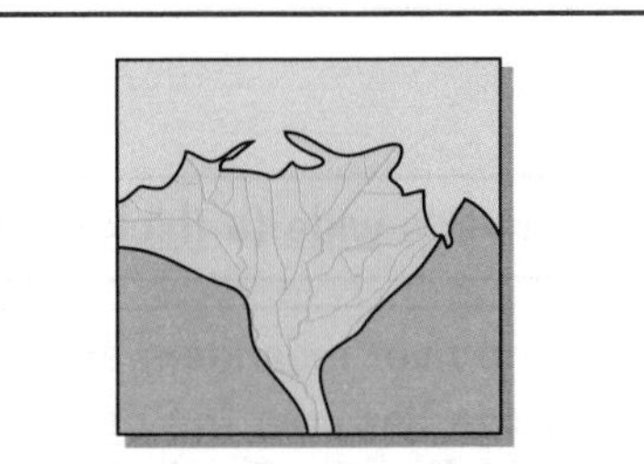

Arcuate — have a rounded shape and lots of distributaries, e.g. the Nile delta.

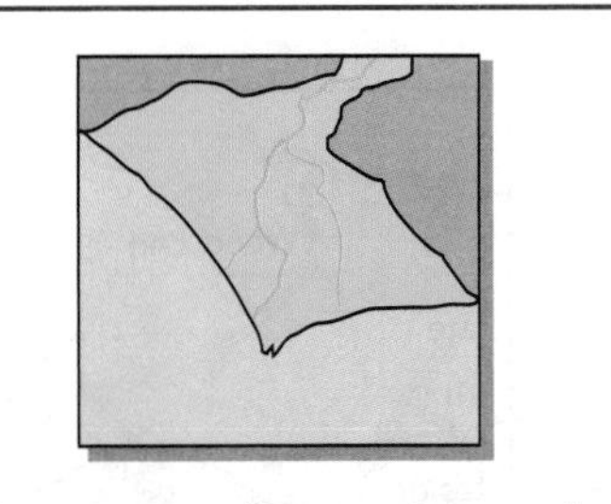

Cuspate — have a triangular shape and few distributaries, e.g. the Tiber delta.

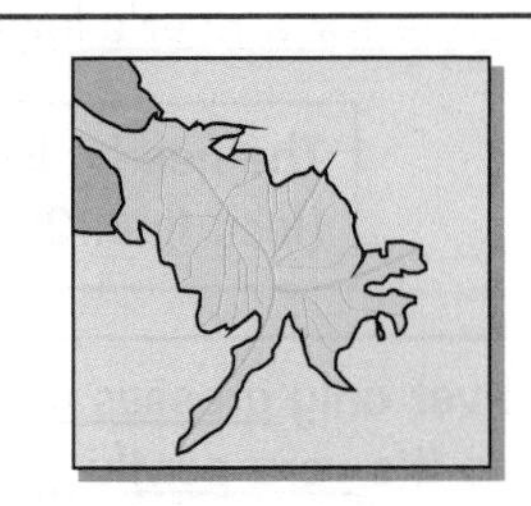

Bird's foot — wait for it... are shaped like a bird's foot, e.g. the Mississippi delta.

Yeah, it looks like a bird's foot — if the bird landed in a blender...

I'll be the first person to admit that these depositional landforms aren't as exciting as waterfalls, but it's still worth knowing about them. The only tricky bits on this page are the names of the types of delta — the rest is a piece of cake.

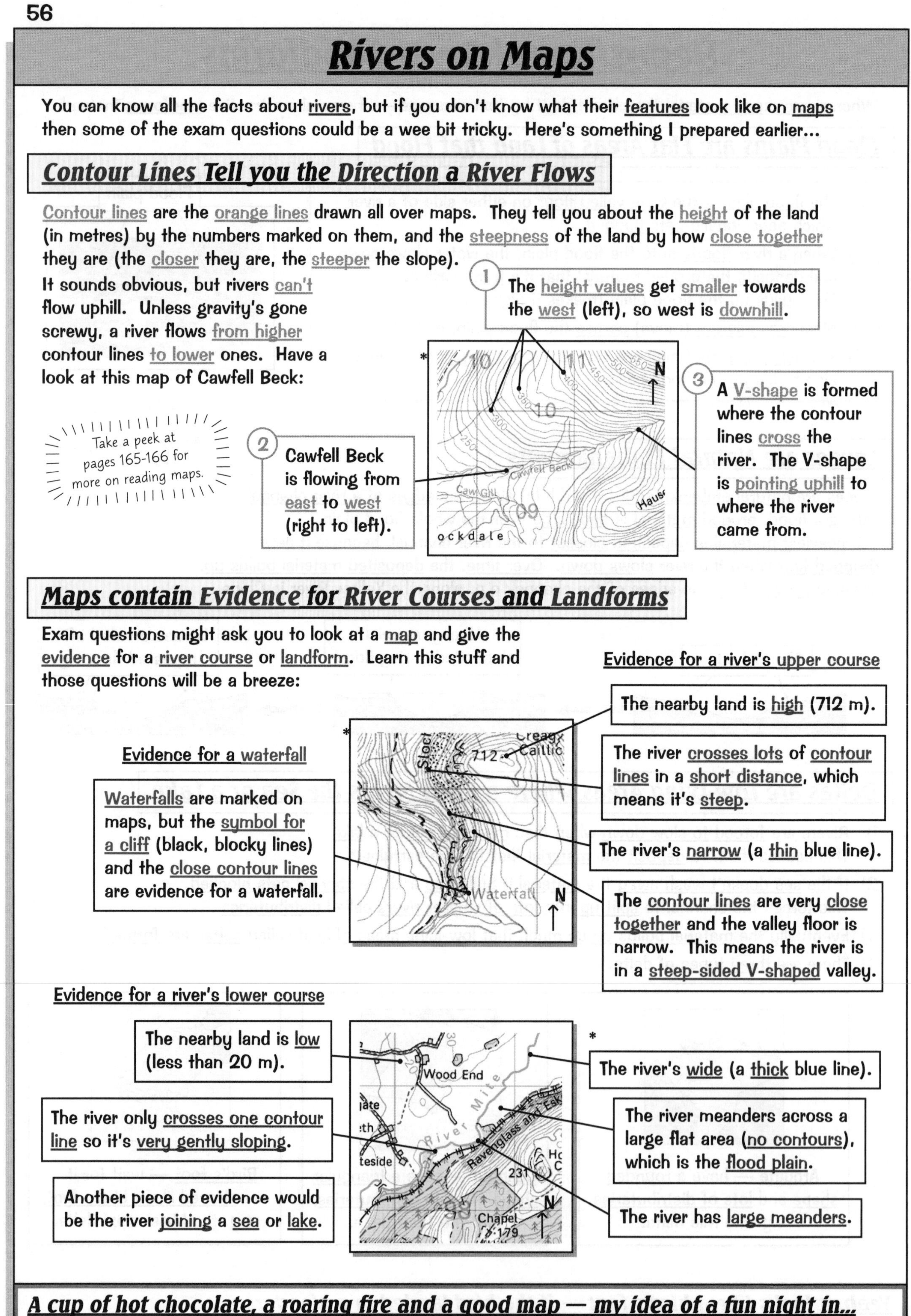

Rivers on Maps

You can know all the facts about rivers, but if you don't know what their features look like on maps then some of the exam questions could be a wee bit tricky. Here's something I prepared earlier...

Contour Lines Tell you the Direction a River Flows

Contour lines are the orange lines drawn all over maps. They tell you about the height of the land (in metres) by the numbers marked on them, and the steepness of the land by how close together they are (the closer they are, the steeper the slope).

It sounds obvious, but rivers can't flow uphill. Unless gravity's gone screwy, a river flows from higher contour lines to lower ones. Have a look at this map of Cawfell Beck:

Take a peek at pages 165-166 for more on reading maps.

1. The height values get smaller towards the west (left), so west is downhill.
2. Cawfell Beck is flowing from east to west (right to left).
3. A V-shape is formed where the contour lines cross the river. The V-shape is pointing uphill to where the river came from.

Maps contain Evidence for River Courses and Landforms

Exam questions might ask you to look at a map and give the evidence for a river course or landform. Learn this stuff and those questions will be a breeze:

Evidence for a river's upper course

- The nearby land is high (712 m).
- The river crosses lots of contour lines in a short distance, which means it's steep.
- The river's narrow (a thin blue line).
- The contour lines are very close together and the valley floor is narrow. This means the river is in a steep-sided V-shaped valley.

Evidence for a waterfall

Waterfalls are marked on maps, but the symbol for a cliff (black, blocky lines) and the close contour lines are evidence for a waterfall.

Evidence for a river's lower course

- The nearby land is low (less than 20 m).
- The river only crosses one contour line so it's very gently sloping.
- Another piece of evidence would be the river joining a sea or lake.
- The river's wide (a thick blue line).
- The river meanders across a large flat area (no contours), which is the flood plain.
- The river has large meanders.

A cup of hot chocolate, a roaring fire and a good map — my idea of a fun night in...

Map questions can be a goldmine of easy marks — all you have to do is say what you see. You just need to understand what the maps are showing, so read this page like there's no tomorrow, then see if you can remember it all.

River Valley — Case Study

I know what you're thinking — we're nearly 10 pages into this section and there's not been a case study yet. Well, thank your lucky stars because I've got a real treat lined up for you — we're off to Glasgow...

The River Clyde Flows Through Scotland

1) The River Clyde is about 160 km long.
2) Its source is in the Southern Uplands region of Scotland and the river flows north-west through Motherwell and Glasgow.
3) The mouth of the River Clyde is an estuary on the west coast of Scotland.
4) Here are some of the features and landforms in the valley that the River Clyde flows through:

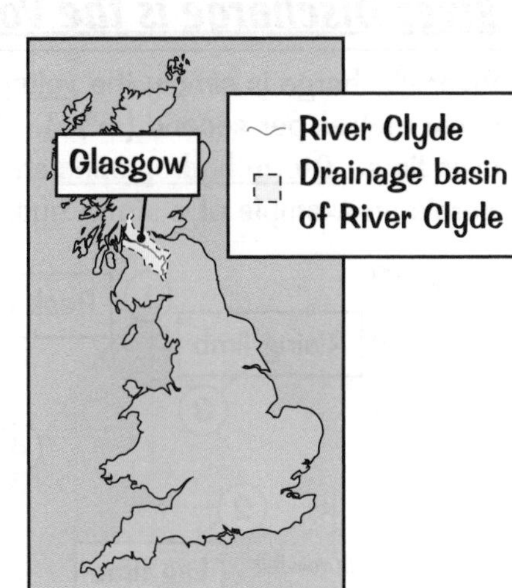

An estuary is the mouth of a river that joins the sea.

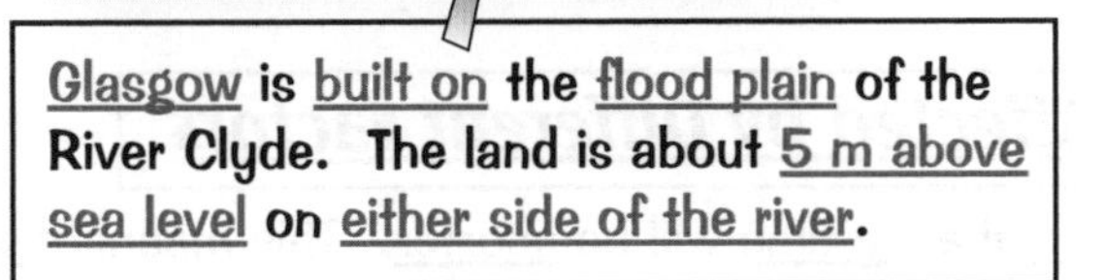

The river's estuary is about 34 km west of Glasgow — the estuary is about 3 km wide. The river joins the Firth of Clyde, which eventually becomes the Irish Sea.

Glasgow is built on the flood plain of the River Clyde. The land is about 5 m above sea level on either side of the river.

There's an ox-bow lake forming near the village of Uddingston.

The Falls of Clyde are four waterfalls near Lanark. The highest fall is Corra Linn — it's about 27 m high. There's also a gorge along this part of the river, formed by the waterfalls retreating.

Glasgow

Motherwell

Lanark

direction of flow

Crawford

The river meanders between Motherwell and Glasgow.

There are interlocking spurs at Crawford. The spurs are between 300 and 500 m high.

The source of the river is in the Lowther Hills — two tributaries (Daer Water and Portail Water) come together to form the River Clyde.

Jocky and Clyde — doesn't really have the same ring to it...

What a lovely case study to ease you in with. You don't have to thank me, but it would be nice. The important bits of this case study are the landforms, place names and measurements, so learn 'em. Job done. Let's move on...

River Discharge

We've not really talked much about the actual water in a river. Well, all that's about to change — hooray.

River Discharge is the Volume of Water Flowing in a River

River discharge is simply the volume of water that flows in a river per second. It's measured in cumecs — cubic metres per second (m^3/s). Hydrographs show how the discharge at a certain point in a river changes over time. Storm hydrographs show the changes in river discharge around the time of a storm. Here's an example of a storm hydrograph:

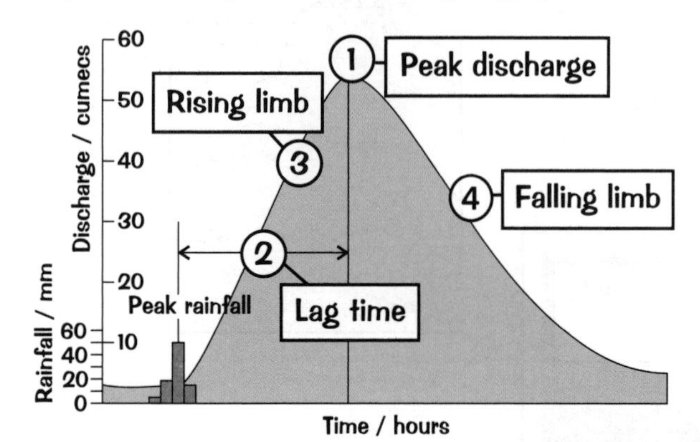

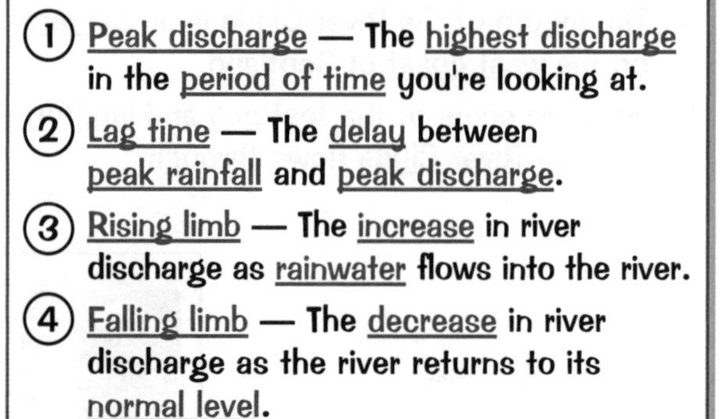

1. Peak discharge — The highest discharge in the period of time you're looking at.
2. Lag time — The delay between peak rainfall and peak discharge.
3. Rising limb — The increase in river discharge as rainwater flows into the river.
4. Falling limb — The decrease in river discharge as the river returns to its normal level.

Lag time happens because most rainwater doesn't land directly in the river channel — there's a delay as rainwater gets to the channel. It gets there by flowing quickly overland (called surface runoff, or just runoff), or by soaking into the ground (called infiltration) and flowing slowly underground.

River Discharge is Affected by Different Factors

If more water flows as runoff the lag time will be shorter. This means discharge will increase and the hydrograph will be steeper because more water gets to the river in a shorter space of time. Here are a few factors that affect discharge and the shape of the hydrograph:

Factors that increase discharge and make the hydrograph steeper	Factors that decrease discharge and make the hydrograph gentler
High rainfall causes more runoff and a shorter lag time.	Low rainfall causes less runoff and a longer lag time.
Intense rainfall causes more runoff and a shorter lag time.	Light rainfall causes less runoff and a longer lag time.
Impermeable rock — water can't infiltrate into the rock so there's more runoff and a shorter lag time.	Permeable rock — water infiltrates through pore spaces in permeable rock, so there's less runoff and a longer lag time.
Hot, dry conditions and freezing conditions both result in hard ground, so there's more runoff and a shorter lag time.	Mild conditions result in soft ground so water can infiltrate. There's less runoff and a longer lag time.
Previously wet conditions — water can't infiltrate into saturated soil so there's more runoff and a shorter lag time.	Previously dry conditions — water can infiltrate into dry soil so there's less runoff and a longer lag time.
Steep slopes cause more runoff and a shorter lag time.	Gentle slopes cause less runoff and a longer lag time.
Less vegetation means less water is intercepted and evaporates so more water reaches the channel. Throughflow isn't slowed down by roots, so there's a shorter lag time.	More vegetation means more water is intercepted and evaporates, so less water reaches the channel. Throughflow is also slowed down by roots, so there's a longer lag time.

URBAN AREAS have drainage systems and they're covered with impermeable materials — these increase discharge so hydrographs for rivers in urban areas are steep.

RURAL AREAS have more vegetation, which decreases discharge. There are also more reservoirs in rural areas — they store water and release it slowly, decreasing discharge in the river below. This means hydrographs for rivers in rural areas are more gently sloping.

Revision lag time — the time between starting and getting bored...

There are loads of different factors that can affect the discharge of a river and hydrographs are a helpful way of understanding what's going on. They sound like something from the future, but the excitement ends there I'm afraid.

Flooding

Flooding happens when the level of a river gets so high that it spills over its banks.

Rivers Flood due to Physical Factors

The river level increases when the discharge increases because a high discharge means there's more water in the channel. This means the factors that increase discharge can also cause flooding:

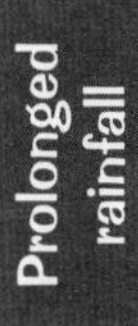

After a long period of rain, the soil becomes saturated. Any further rainfall can't infiltrate, which increases runoff into rivers. This increases discharge quickly, which can cause a flood.

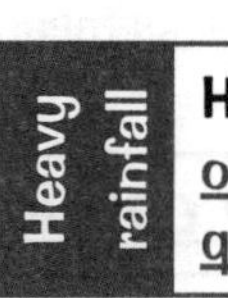

Heavy rainfall means there's a lot of runoff. This increases discharge quickly, which can cause a flood.

Relief

Relief is how the height of the land changes. If a river is in a steep-sided valley, water will reach the river channel much faster because water flows more quickly on steeper slopes. This increases discharge quickly, which can cause a flood.

Snowmelt

When a lot of snow or ice melts it means that a lot of water goes into a river in a short space of time. This increases discharge quickly, which can cause a flood.

Geology

When a river is in an area of permeable rock (e.g. limestone), more water percolates into the rock instead of flowing on the surface. This means there's less runoff, so the risk of flooding is lower. When a river is in an area of impermeable rock (e.g. clay), water doesn't percolate into the rock but flows on the surface. This means there's more runoff, so the risk of flooding is higher.

Rivers also Flood because of Human Factors

Here are a couple of examples of how human actions can make flooding more frequent and more severe:

Deforestation

Trees intercept rainwater on their leaves, which then evaporates. Trees also take up water from the ground and store it. This means cutting down trees increases the volume of water that reaches the river channel, which increases discharge and makes flooding more likely.

Building construction

Buildings are often made from impermeable materials, e.g. concrete, and they're surrounded by roads made from tarmac (also impermeable). Impermeable surfaces increase runoff and drains quickly take runoff to rivers. This increases discharge quickly, which can cause a flood.

River Flooding in the UK Appears to be Happening More Often

Some rivers in the UK have been flooding more frequently over the last 20 years. For example, the River Ouse in Yorkshire reached a high water level 29 times between 1966 and 1986. But between 1987 and 2007 it reached the same level 80 times.

The table shows the locations and dates of some of the big floods that have happened in the UK since 1988.

Year	Rivers	Places affected
1988	Kenwyn	Cornwall
1990	Severn	Gloucestershire
1994	Lavant, Clyde	West Sussex, Glasgow
1998	Severn, Trent, Wye	The Midlands, Mid and South Wales
2000	Ouse, Alyn	Yorkshire, North Wales
2004	Valency	Cornwall
2005	Eden	Cumbria and North Yorkshire
2007	Many	Many parts of the UK
2008	Severn	South Midlands

Lots of water causes flooding — this mainly applies to rivers and toddlers...

If you're having a bath and you leave the taps on the water will eventually go over the sides. It's the same with rivers — they've got a limit to the volume of water they can hold. Please don't experiment by flooding the bathroom though.

Flood Management

Floods can be devastating, but there are a number of different strategies to stop them or lessen the blow.

Floods have Some Serious Impacts

1) Floods have many impacts, but the most serious ones are that people are killed by flood waters, buildings are damaged or destroyed and jobs are lost because of damage to premises and equipment.
2) The effects of flooding are worse in poorer countries than richer countries because there's less money to spend on flood protection and to help people after a flood. Also, more people live and work in areas that are likely to flood and poorer transport links mean it's more difficult to get help to places that have been affected.

Hard and Soft Engineering can Reduce the Risk of Flooding or its Effects

Hard engineering — man-made structures built to control the flow of rivers and reduce flooding.

Soft engineering — schemes set up using knowledge of a river and its processes to reduce the effects of flooding.

Strategy	What it is	Benefits	Disadvantages
Dams and reservoirs	Dams (huge walls) are built across the rivers, usually in the upper course. A reservoir (artificial lake) is formed behind the dam.	Reservoirs store water and release it slowly, which reduces the risk of flooding. The water in the reservoir is used as drinking water and can be used to generate hydroelectric power (HEP). Reservoirs are also attractive and can be used for recreation.	Dams are very expensive to build. Creating a reservoir can flood existing settlements. Eroded material is deposited in the reservoir and not along the river's natural course making farmland downstream less fertile.
Channel straightening	The river's course is straightened — meanders are cut out by building artificial straight channels.	Water moves out of the area more quickly because it doesn't travel as far — reducing the risk of flooding.	Flooding may happen downstream of the straightened channel instead, as flood water is carried there faster.
Man-made levees	Man-made embankments along both sides of the river.	The embankments mean that the river channel can hold more water, which reduces the risk of flooding. They're also quite cheap.	If the levees fail (break) it can cause catastrophic flooding.
Flood warnings	People are warned about possible flooding through TV, radio, newspapers and the internet.	The impact of flooding is reduced — warnings give people time to move possessions upstairs, put sandbags in position and to evacuate.	Warnings don't stop a flood from happening. People may not hear or have access to warnings (especially in poorer countries where communications are less developed).
Preparation	Buildings are modified to reduce the amount of damage a flood could cause. People make plans for what to do in a flood, e.g. keep a blanket and torch in a handy place.	The impact of flooding is reduced — buildings are less damaged and people know what to do when a flood happens.	Preparation doesn't guarantee safety from a flood and it could give people a false sense of security. It's expensive to modify homes and businesses.
Flood plain zoning	Restrictions prevent building on parts of a flood plain that are likely to be affected by a flood.	The risk of flooding is reduced — impermeable surfaces aren't created, e.g. roads. The impact of flooding is reduced — there aren't any houses or roads to damage.	The expansion of an urban area is limited if there aren't any other suitable building sites. It's no help in areas that have already been built on.
'Do nothing'	No money is spent on new engineering methods or maintaining existing ones. Flooding is a natural process and people should accept the risks of living in an area that's likely to flood.	The river floods, eroded material is deposited on the flood plain, making farmland more fertile.	The risk of flooding and the impacts of flooding aren't reduced. A flood will probably cause a lot of damage.

Some strategies for flood management are more sustainable than others. Sustainable strategies meet the needs of people today (i.e. they reduce flooding), without stopping people in the future getting the things they need. This means not using up resources (e.g. money) or damaging the environment.

Hard engineering strategies are usually less sustainable than soft engineering strategies because they generally cost more to build and maintain, and they damage the environment more.

No, river straightening isn't done with a gigantic pair of hot ceramic plates...

Flooding can be a nightmare, especially if you live in a poorer country. But, as luck would have it, there are plenty of strategies to reduce the impacts. What's less lucky is the fact that they might come up in the exam, so get learning.

Flooding — Case Studies

Everyone's favourite part of geography — the inevitable case studies. Put your learning hats on...

Rich and Poor parts of the World are Affected Differently by Flooding

The effects of floods and the responses to them are different in different parts of the world. A lot depends on how wealthy the part of the world is. Learn the following case studies — you might have to compare two floods like these in your exam:

	Flood in a rich part of the world:	Flood in a poor part of the world:
	Place: Carlisle, England Date: 8th January, 2005 River: Eden	Place: South Asia (Bangladesh and India) Date: July and August, 2007 Rivers: Brahmaputra and Ganges
Causes	• Heavy rainfall — 200 mm of rain fell in 36 hours. The continuous rainfall saturated the soil, increasing runoff into the River Eden. • Carlisle is a large urban area — impermeable materials like concrete increased runoff. • This caused the discharge of the River Eden to reach 1520 cumecs (its average is 52 cumecs).	• Heavy rainfall — in one region, 900 mm of rain fell in July. • The continuous rainfall saturated the soil, increasing runoff into rivers. • Melting snow from glaciers in the Himalayan mountains increased the discharge of the Brahmaputra river. • The peak discharge of both rivers happened at the same time, which increased discharge downstream.
Primary effects	• 3 deaths. • 350 businesses were shut down. • Some roads and bridges were damaged. • Rivers were polluted with rubbish and sewage.	• Over 2000 deaths. • Many factories closed and lots of livestock were killed. • 112 000 houses were destroyed in India. • Rivers were polluted with rubbish and sewage.
Secondary effects	• Around 3000 people were made homeless. • Children lost out on education — one school was closed for months. • Stress-related illnesses increased after the floods. • Around 3000 jobs were at risk in businesses affected by floods.	• Around 25 million people were made homeless. • Children lost out on education — around 4000 schools were affected by the floods. • Around 100 000 people caught water-borne diseases like dysentery and diarrhoea. • Flooded fields reduced basmati rice yields — prices rose 10%. • Many farmers and factory workers became unemployed.
Flood management	• The Environment Agency monitors river levels and issues flood warnings to the public, local authorities and the media. • The local council distributes sandbags when flood warnings have been issued. • There are man-made levees along the river to help prevent flooding.	• Bangladesh has a Flood Forecasting and Warning System (FFWS) with 85 flood monitoring stations. Flood warnings can be issued up to 72 hours before a flood occurs, but the warnings don't reach many rural communities. • There are around 6000 km of man-made levees to prevent flooding in Bangladesh, but they're easily eroded and aren't properly maintained so are often breached by flood waters.
Immediate response	• People were evacuated from areas that flooded. • Reception centres were opened around Carlisle to provide food and drinks for evacuees. • Temporary accommodation was set up for the people made homeless.	• Many people didn't evacuate from areas that flooded, and blocked transport links slowed down any evacuations that were attempted. • Other governments and international charities distributed food, water and medical aid. Technical equipment like rescue boats were also sent to help people who were stranded.
Long-term response	• Community groups were set up to provide emotional support and to give practical help to people who were affected by the floods. • The Eden and Petteril Flood Alleviation Scheme was completed in 2008 — this involved things like building up flood defence walls and levees on the rivers to prevent flooding.	• International charities have funded the rebuilding of homes and the agriculture and fishing industries. • Some homes have been rebuilt on stilts, so they're less likely to be damaged by future floods.

Flood your mind — with knowledge...

Well, it's pretty clear that floods have different impacts and the responses to them are different when you compare a rich and a poor part of the world. You might be asked to write about a couple of examples in the exam, so why not learn these two...

Managing the UK's Water

You think you can't live without your MP3 player, but believe me, it's water you can't live without.

The Demand for Water is Different Across the UK

In the UK, the places with a good supply of water aren't the same as the places with the highest demand:

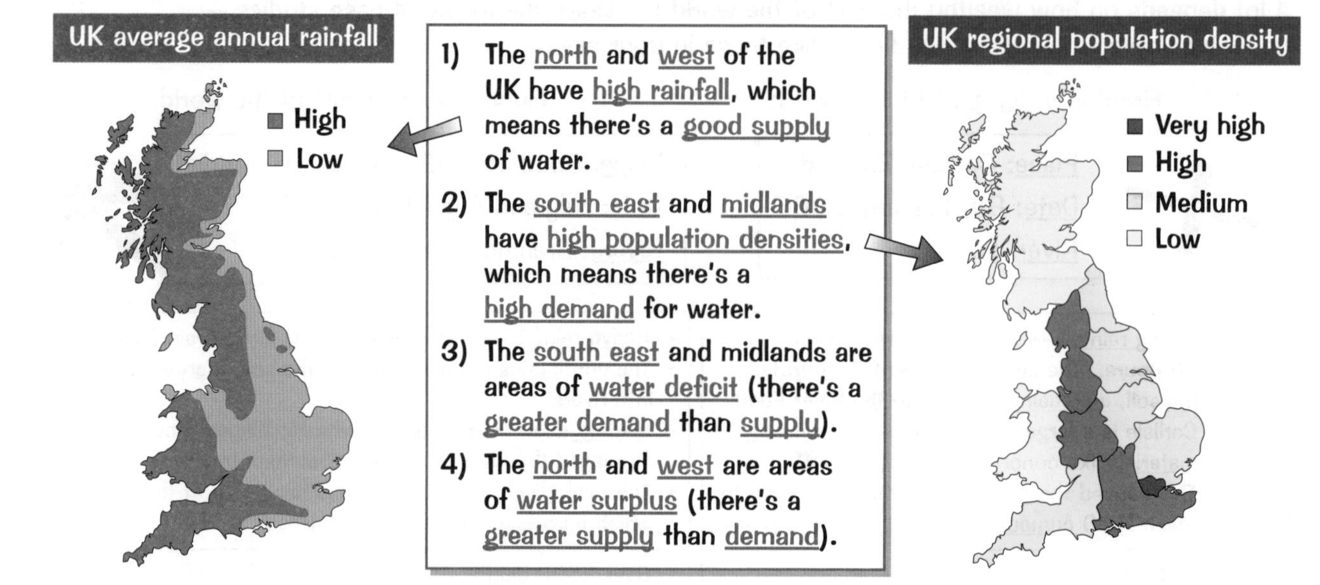

1) The north and west of the UK have high rainfall, which means there's a good supply of water.
2) The south east and midlands have high population densities, which means there's a high demand for water.
3) The south east and midlands are areas of water deficit (there's a greater demand than supply).
4) The north and west are areas of water surplus (there's a greater supply than demand).

The demand for water in the UK is increasing:

1) Over the past 25 years, the amount of water used by people in the UK has gone up by about 50%.
2) The UK population is predicted to increase by around 10 million people over the next 20 years.

The UK needs to Manage its Supply of Water...

1) One way to deal with the supply and demand problem is to transfer water from areas of surplus to areas of deficit. For example, Birmingham (an area of deficit) is supplied with water from the middle of Wales (an area of surplus).
2) Water transfer can cause a variety of issues:
   - The dams and aqueducts (bridges used to transport water) that are needed are expensive.
   - It could affect the wildlife that lives in the rivers, e.g. fish migration patterns could be disrupted by dam building.
   - There might be political issues, e.g. people may not want their water given to another country.
3) Another way to increase water supplies in deficit areas is to build more reservoirs to store more water. However, building a reservoir can involve flooding settlements and relocating people.
4) Fixing leaky pipes would mean less water is lost during transfer. For example, millions of litres of water are lost everyday through leaky pipes around London — fixing leaky pipes would save some of this.

...and Reduce its Demands for Water

1) People can reduce the amount of water that they use at home, e.g. by taking showers instead of baths, running washing machines only when they're full and by using hosepipes less.
2) Water companies want people to have water meters installed — meters are used to charge people for the exact volume of water that they use. People with water meters are more likely to be careful with the amount of water they use — they're paying for every drop.

I'd like a glass of water please — hang on, I'll just nip up to Scotland...

After reading this section you probably won't be leaving the tap on while you brush your teeth again. The UK isn't a desert by any means, but there's an increasing demand for water, which means that supplies have to be managed.

UK Reservoir — Case Study

Joy of joys, another case study. It's about a reservoir in the UK that supplies water to a lot of people...

Rutland Water is a Reservoir in the East Midlands

1) The dam was built and Rutland Water was created during the 1970s.
2) The reservoir covers a 12 km^2 area and it's filled with water from two rivers — the River Welland and the River Nene.
3) Rutland Water was designed to supply the East Midlands with more water — enough to cope with rapid population growth in places like Peterborough.
4) Areas around the reservoir are also used as a nature reserve and for recreation.

Here are some of the economic, social and environmental impacts of Rutland Water:

Rutland Water

Peterborough

Economic

- The reservoir boosts the local economy — it's a popular tourist attraction because of the wildlife and recreation facilities.
- Around 6 km^2 of land was flooded to create the reservoir. This included farmland, so some farmers lost their livelihoods.

Social

- Lots of recreational activities take place on and around the reservoir, e.g. sailing, windsurfing, birdwatching and cycling.
- Many jobs have been created to build and maintain the reservoir, and to run the nature reserve and recreational activities.
- Schools use the reservoir for educational visits.
- Two villages were demolished to make way for the reservoir.

Environmental

- Rutland Water is a Site of Special Scientific Interest (SSSI) — an area where wildlife is protected.
- Hundreds of species of birds live around the reservoir and tens of thousands of waterfowl (birds that live on or near water) come to Rutland Water over the winter.
- A variety of habitats are found around the reservoir, e.g. marshes, mudflats and lagoons. This means lots of different organisms live in or around the reservoir.
- Ospreys (fish-eating birds of prey that were extinct in Britain) have been reintroduced to central England by the Rutland Osprey Project at the reservoir.
- A large area of land was flooded to create the reservoir, which destroyed some habitats.

Rutland Water has to be Managed Sustainably

The supply of water from the reservoir has to be sustainable. This means that people should be able to get all the water they need today, without stopping people in the future from having enough water.

Basically, people today can't deplete the water supply or damage the environment too much, or the supply won't be the same in the future. To use the reservoir in a sustainable way, people can only take out as much water as is replaced by the rivers that supply it. That way, the supply will stay the same for the future.

Don't get stuck in a Rut-land while doing your revision...

Congratulations, another section under your belt — your prize awaits you on the next page. OK, so it's a revision summary, but the real prize is the satisfaction of knowing everything about rivers. I hope that sounded convincing...

Revision Summary for Section 5

Wat-er load of fun that was. Now it's time to see how much information your brain has soaked up. I think you'll be surprised — I reckon about 16 litres of knowledge has been taken in. Have a go at the questions below and then go back over the section to check your answers. If something's not quite right, pore over the page again. Once you can answer everything correctly you're ready to sail away, sail away, sail away... to the next section.

1) What does the hydrological cycle show?
2) Describe three flows and three stores in the hydrological cycle.
3) What is a drainage basin?
4) Name two outputs from a drainage basin system.
5) What happens at a confluence?
6) Describe freeze-thaw weathering.
7) What does a river's long profile show?
8) Describe the cross profile of a river's lower course.
9) Name the river course where vertical erosion is dominant.
10) What's the difference between abrasion and attrition?
11) Name two processes of transportation.
12) When does deposition occur?
13) Where do waterfalls form? Name an example.
14) How is a gorge formed?
15) Why do rivers have to wind around interlocking spurs?
16) Where is the current fastest on a meander?
17) Name the landform created when a meander is cut off by deposition.
18) What is a flood plain?
19) Describe how levees are formed.
20) Name one type of delta and describe what it looks like.
21) What do the contour lines on a map show?
22) Give two pieces of map evidence for either a waterfall or a river's lower course.
23) Describe three landforms in the valley of a river you have studied.
24) What is river discharge?
25) How does impermeable rock affect river discharge?
26) Describe two physical factors and one human factor that can increase the risk of floods.
27) Name a hard engineering strategy and describe its benefits.
28) Name a soft engineering strategy and describe its disadvantages.
29) Why are hard engineering strategies usually less sustainable than soft engineering strategies?
30) a) Name a flood that you have studied and state when it occurred.
    b) Describe the primary effects of the flood and the long-term responses to it.
31) Which areas of the UK have a water deficit?
32) Give one potential problem of water transfer.
33) a) Name a reservoir in the UK.
    b) Describe two social impacts and two environmental impacts that the reservoir has had.
34) How can water be taken from a reservoir in a sustainable way?

Ice Levels Over Time

It's time to put your hat, scarf and jumper on, you've arrived at the section that's all about ice...

The Earth has Glacial Periods and Interglacial Periods

1) The Earth goes through cold periods which last for millions of years called ice ages. During ice ages, large masses of ice cover parts of the Earth's surface.
2) The last ice age was the Pleistocene that began around 2.6 million years ago.
3) During ice ages there are cooler periods called glacial periods when the ice advances to cover more of the Earth's surface. Each one lasts for about 100 000 years.
4) In between the glacial periods are warmer periods called interglacial periods when the ice retreats to cover less of the Earth's surface. Each one lasts around 10 000 years.
5) The last glacial period began around 100 000 years ago and ended around 10 000 years ago.

Ice Covered Much More of the Earth's Surface 20 000 Years Ago

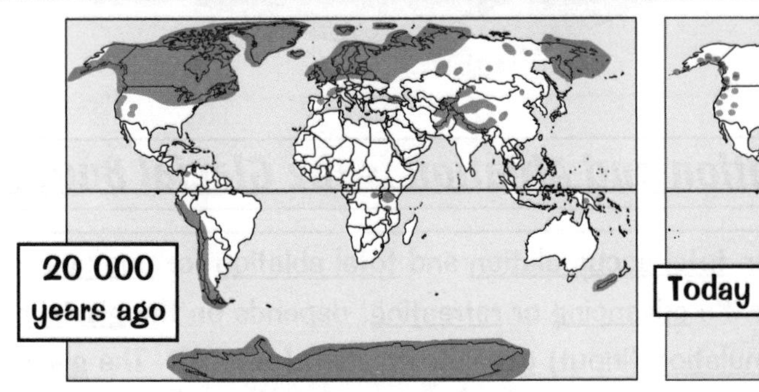

Today

= Ice

1) Since the beginning of the Pleistocene there have been permanent ice sheets on Greenland and Antarctica. Ice has also covered other parts of the world during the colder glacial periods.
2) Ice covered a lot more of the land around 20 000 years ago (during the last glacial period) — over 30% of the Earth's land surface was covered by ice, including nearly all of the UK.
3) We're currently in an interglacial period that began around 10 000 years ago. Today about 10% of the Earth's land surface is covered by ice — the only ice sheets are the ones on Greenland and Antarctica.

Ice sheets are huge masses of ice that cover whole continents. Glaciers are masses of ice which fill valleys and hollows.

Evidence of Changing Temperature Comes From Three Main Sources

CHEMICAL EVIDENCE

The chemical composition of ice and marine sediments change as temperature changes, so they can be used to work out how global temperature has changed in the past. Ice and sediments build up over thousands of years so samples taken at different depths show the temperature over thousands of years. The records show a pattern of increasing and decreasing temperature, which caused the ice to advance and retreat.

GEOLOGICAL EVIDENCE

Some landforms we can see today were created by glaciers in the past (see p. 68). This shows that some areas that aren't covered in ice today were covered in the past, which means temperatures were lower.

FOSSIL EVIDENCE

The remains of some organisms are preserved when they die, creating fossils. Fossils show the distribution of plants and animals that are adapted to warm or cold climates at different times in the past. From this we can tell which areas were warmer or colder in the past.

Pleistocene — comes in many colours and gets stuck in the carpet...

Don't take your warm clothes off yet — there's lots more ice to come. For now, see if you can remember what an ice age is, what a glacial period is, when the last glacial period was, and how far the ice sheets spread in that period.

Glacial Budget

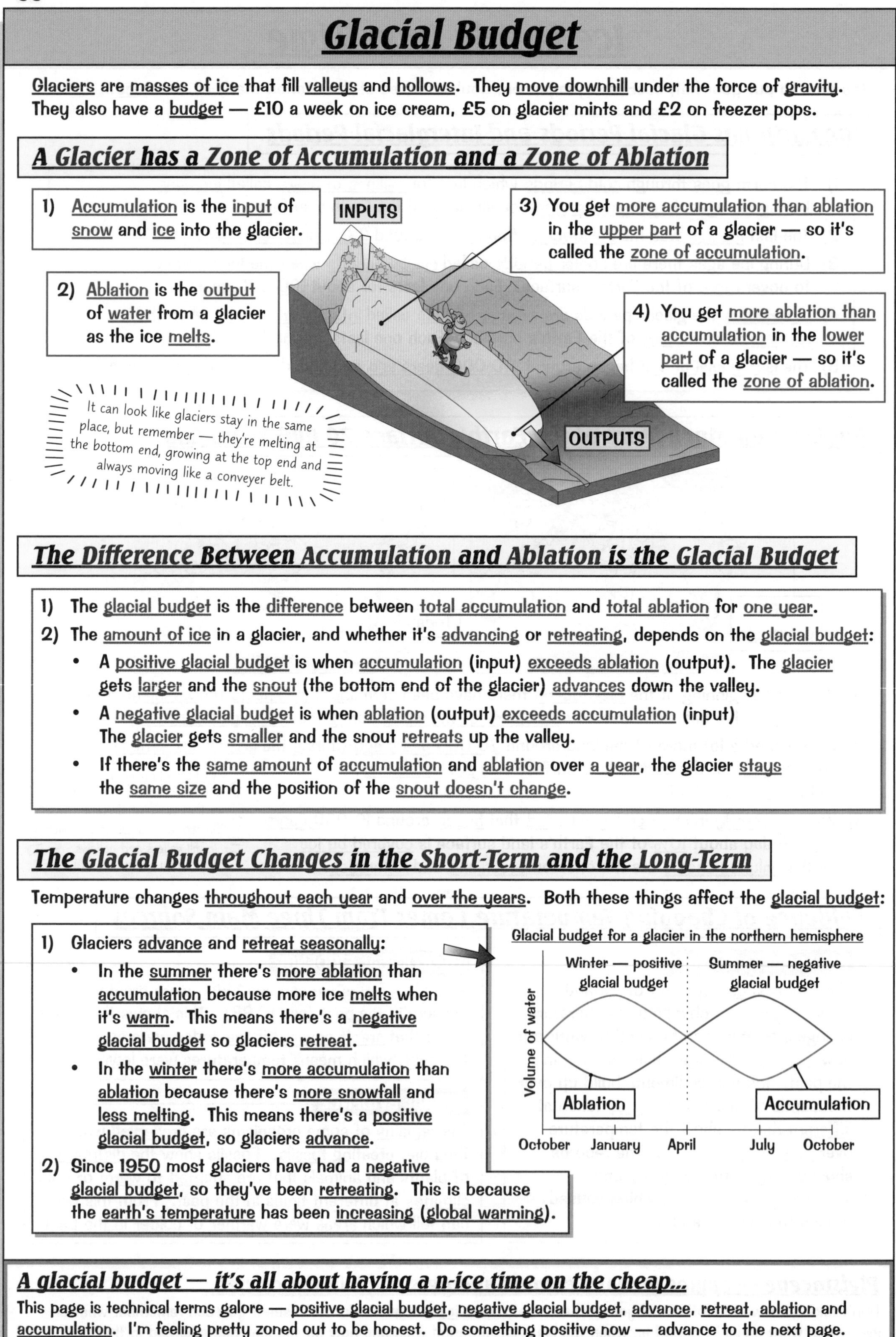

Glaciers are masses of ice that fill valleys and hollows. They move downhill under the force of gravity. They also have a budget — £10 a week on ice cream, £5 on glacier mints and £2 on freezer pops.

A Glacier has a Zone of Accumulation and a Zone of Ablation

1) Accumulation is the input of snow and ice into the glacier.
2) Ablation is the output of water from a glacier as the ice melts.
3) You get more accumulation than ablation in the upper part of a glacier — so it's called the zone of accumulation.
4) You get more ablation than accumulation in the lower part of a glacier — so it's called the zone of ablation.

It can look like glaciers stay in the same place, but remember — they're melting at the bottom end, growing at the top end and always moving like a conveyer belt.

The Difference Between Accumulation and Ablation is the Glacial Budget

1) The glacial budget is the difference between total accumulation and total ablation for one year.
2) The amount of ice in a glacier, and whether it's advancing or retreating, depends on the glacial budget:
   - A positive glacial budget is when accumulation (input) exceeds ablation (output). The glacier gets larger and the snout (the bottom end of the glacier) advances down the valley.
   - A negative glacial budget is when ablation (output) exceeds accumulation (input) The glacier gets smaller and the snout retreats up the valley.
   - If there's the same amount of accumulation and ablation over a year, the glacier stays the same size and the position of the snout doesn't change.

The Glacial Budget Changes in the Short-Term and the Long-Term

Temperature changes throughout each year and over the years. Both these things affect the glacial budget:

1) Glaciers advance and retreat seasonally:
   - In the summer there's more ablation than accumulation because more ice melts when it's warm. This means there's a negative glacial budget so glaciers retreat.
   - In the winter there's more accumulation than ablation because there's more snowfall and less melting. This means there's a positive glacial budget, so glaciers advance.
2) Since 1950 most glaciers have had a negative glacial budget, so they've been retreating. This is because the earth's temperature has been increasing (global warming).

A glacial budget — it's all about having a n-ice time on the cheap...

This page is technical terms galore — positive glacial budget, negative glacial budget, advance, retreat, ablation and accumulation. I'm feeling pretty zoned out to be honest. Do something positive now — advance to the next page.

Glacier — Case Study

If you fancy a sight-seeing trip to a glacier you'd better book it quick. Most glaciers are getting smaller, some by up to a few metres each year — and some are in serious danger of disappearing completely.

The Rhône Glacier is Retreating

1) The Rhône Glacier is in the Swiss Alps.
2) It's currently about 7.8 km long.
3) Like most of the glaciers in the world, it's been retreating since the 19th century.

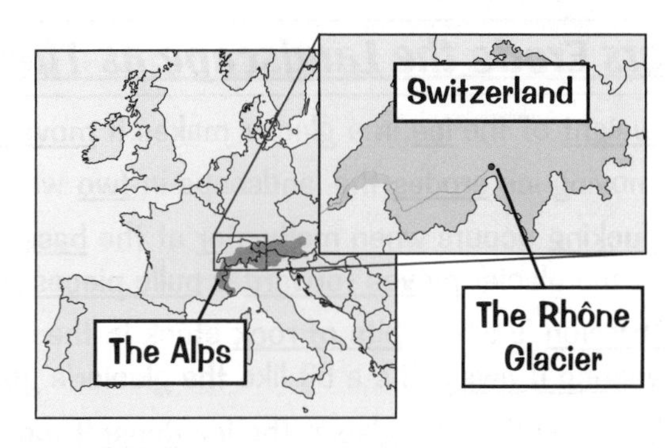

Evidence of Glacial Retreat comes from Various Sources

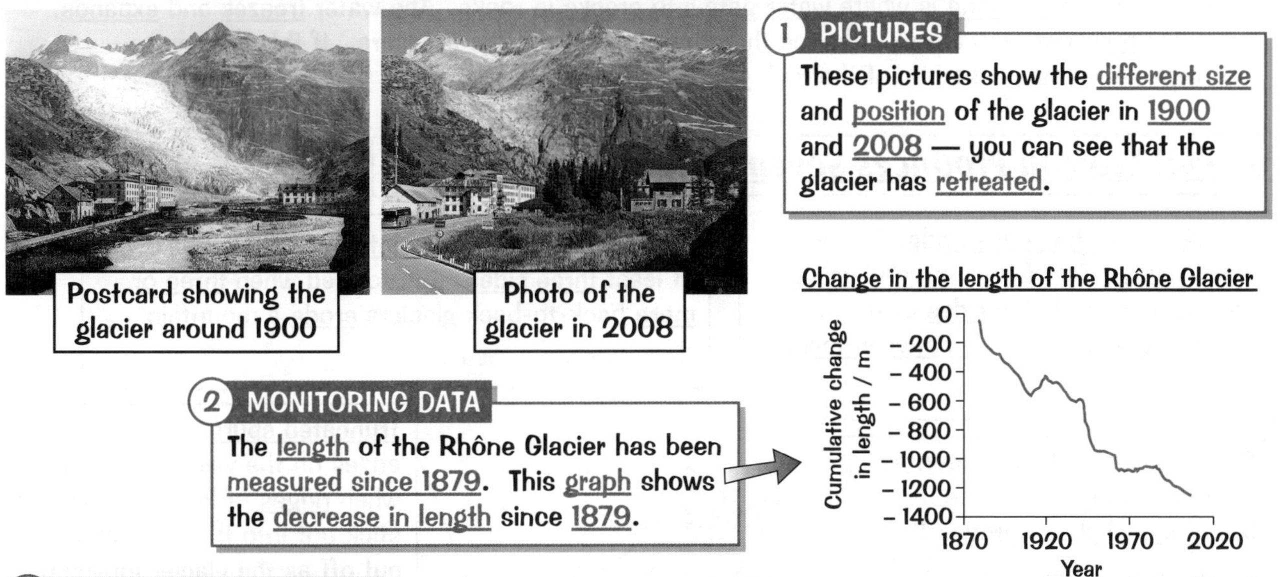

Postcard showing the glacier around 1900

Photo of the glacier in 2008

1 PICTURES

These pictures show the different size and position of the glacier in 1900 and 2008 — you can see that the glacier has retreated.

2 MONITORING DATA

The length of the Rhône Glacier has been measured since 1879. This graph shows the decrease in length since 1879.

3 AMOUNT OF MELTWATER

As the glacier retreats it produces more meltwater. The meltwater has formed a new lake in front of the glacier which has been increasing in size. This shows the glacier has been melting more rapidly.

Global Warming is the Main Cause of Glacial Retreat

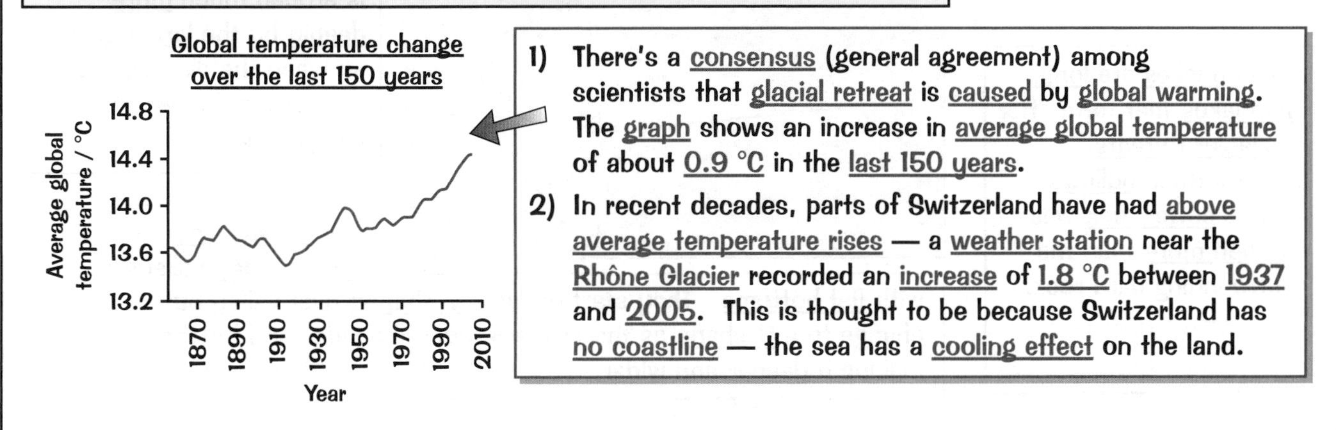

1) There's a consensus (general agreement) among scientists that glacial retreat is caused by global warming. The graph shows an increase in average global temperature of about 0.9 °C in the last 150 years.
2) In recent decades, parts of Switzerland have had above average temperature rises — a weather station near the Rhône Glacier recorded an increase of 1.8 °C between 1937 and 2005. This is thought to be because Switzerland has no coastline — the sea has a cooling effect on the land.

Glacial retreat — where ice cubes go for some R&R...

It might sound like a fairly obvious answer, but the evidence for glacial retreat is that the glaciers are getting smaller. Sadly, that might be a little bit too simple for an exam answer — good thing there's a ready-made case study right here.

Glacial Erosion

It might not seem like glaciers do much — after all they're just large blocks of ice sitting around — but they actually cause rather a lot of erosion and have a massive effect on the landscape around us.

Glaciers Erode the Landscape as They Move

1) The weight of the ice in a glacier makes it move downhill (advance), eroding the landscape as it goes.
2) The moving ice erodes the landscape in two ways:
   - Plucking occurs when meltwater at the base, back or sides of a glacier freezes onto the rock. As the glacier moves forward it pulls pieces of rock out.
   - Abrasion is where bits of rock stuck in the ice grind against the rock below the glacier, wearing it away (it's a bit like the glacier's got sandpaper on the bottom of it).
3) At the top end of the glacier the ice doesn't move in a straight line — it moves in a circular motion called rotational slip. This can erode hollows in the landscape and deepen them into bowl shapes.
4) The rock above glaciers is also weathered (broken down where it is) by the conditions around glaciers. Freeze-thaw weathering is where water gets into cracks in rocks. The water freezes and expands, putting pressure on the rock. The ice then thaws, releasing the pressure. If this process is repeated it can make bits of the rock fall off.

Glacial Erosion Produces Seven Different Landforms

An arête is a steep-sided ridge formed when two glaciers flow in parallel valleys. The glaciers erode the sides of the valleys, which sharpens the ridge between them. (E.g. Striding Edge, Lake District)

A pyramidal peak is a pointed mountain peak with at least three sides. It's formed when three or more back-to-back glaciers erode a mountain. (E.g. Snowdon, Wales)

Corries begin as hollows containing a small glacier. As the ice moves by rotational slip, it erodes the hollow into a steep-sided, armchair shape with a lip at the bottom end. When the ice melts it can leave a small circular lake called a tarn. (E.g. Red Tarn, Lake District)

Truncated spurs are cliff-like edges on the valley side formed when ridges of land (spurs) that stick out into the main valley are cut off as the glacier moves past.

Hanging valleys are valleys formed by smaller glaciers (called tributary glaciers) that flow into the main glacier. The glacial trough is eroded much more deeply by the larger glacier, so when the glaciers melt the valleys are left at a higher level.

Ribbon lakes are long, thin lakes that form after a glacier retreats. They form in hollows where softer rock was eroded more than the surrounding hard rock. (E.g. Windermere, Lake District)

Glacial troughs are steep-sided valleys with flat bottoms. They start off as a V-shaped river valley but change to a U-shape as the glacier erodes the sides and bottom, making it deeper and wider. (E.g. Nant Ffrancon, Snowdonia)

Truncated spurs — when Tottenham finish their match early...

Knowing these erosional landforms will help you sound clever next time you're up a mountain... it'll also help in your quest for a GCSE in geography. Spotting them on a map is a useful skill too — see page 70 for help with that.

Glacial Transport and Deposition

Glaciers are a bit like small children — they do an excellent job of damaging their surroundings and then they leave lots of mess behind. Problem is, no-one's going to clean up after them, the mucky pups.

Glaciers Transport and Deposit Material

1) Glaciers can move material (such as rocks and earth) over very large distances — this is called transportation.
2) The material is frozen in the glacier, carried on its surface, or pushed in front of it. It's called bulldozing when the ice pushes loose material in front of it.
3) When the ice carrying the material melts, the material is dropped on the valley floor — this is called deposition. It also occurs when the ice is overloaded with material.
4) The dropped material makes landforms such as moraines and drumlins (see below).
5) Glacial deposits aren't sorted by weight like river deposits — rocks of all shapes and sizes are mixed up together.

Glaciers Deposit Material as Different Types of Moraine

Moraines are landforms made out of material dropped by a glacier as it melts. There are four different types, depending on their position:

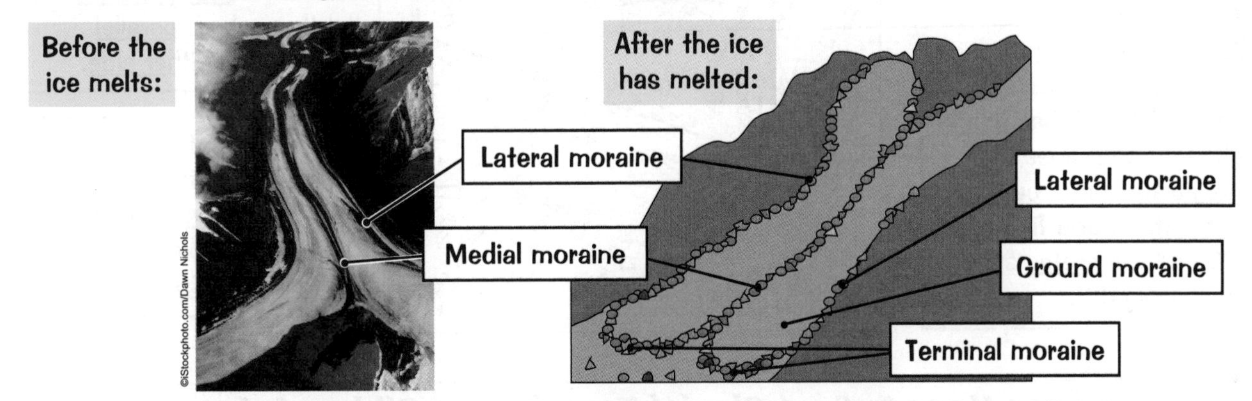

1) Lateral moraine is a long mound of material deposited where the side of the glacier was.
2) Medial moraine is a long mound of material deposited in the centre of a valley where two glaciers met (the two lateral moraines join together).
3) Terminal moraine builds up at the snout of the glacier when it remains stationary. It's deposited as semicircular mounds.
4) Ground moraine is a thin layer of material deposited over a large area as a glacier melts.

Material can also be Deposited as Drumlins

1) Drumlins are elongated hills of glacial deposits — the largest ones can be over 1000 m long, 500 m wide and 50 m high.
2) They're round, blunt and steep at the upstream end, and tapered, pointed and gently sloping at the downstream end.
3) An example of where drumlins can be found is the Ribble Valley, Lancashire.

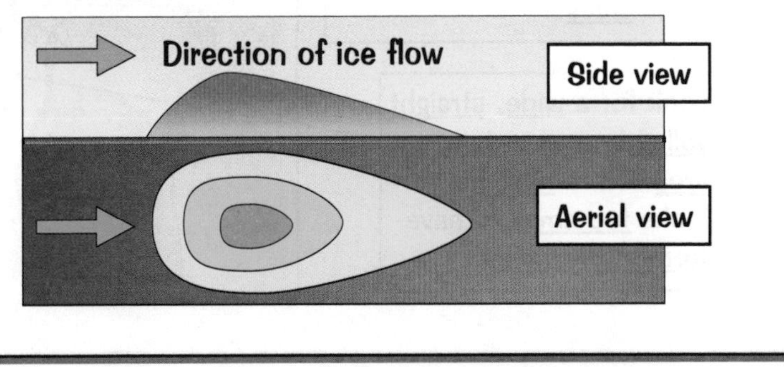

The glacial weather forecast — cold in most places, with mor-raine on the way...

Check you know the differences between the four different types of moraine — try drawing a simple version of the diagram. Now can I have a drumlin roll for the next page... it's all about maps. All geographers love maps. Fact.

Glacial Landforms on Maps

Spotting glacial landforms on OS® maps is an important life skill — no really, it is. It's also a pretty important skill for budding geographers. For a bit more on using maps, turn to pages 167 and 168.

Use Contour Lines to Spot Pyramidal Peaks, Corries and Arêtes on a Map

Contour lines are the orange lines drawn all over maps. They tell you about the height of the land by the numbers marked on them, and the steepness of the land by how close together the lines are (the closer they are, the steeper the slope). Here are a few tips on how to spot pyramidal peaks, arêtes and corries on a map:

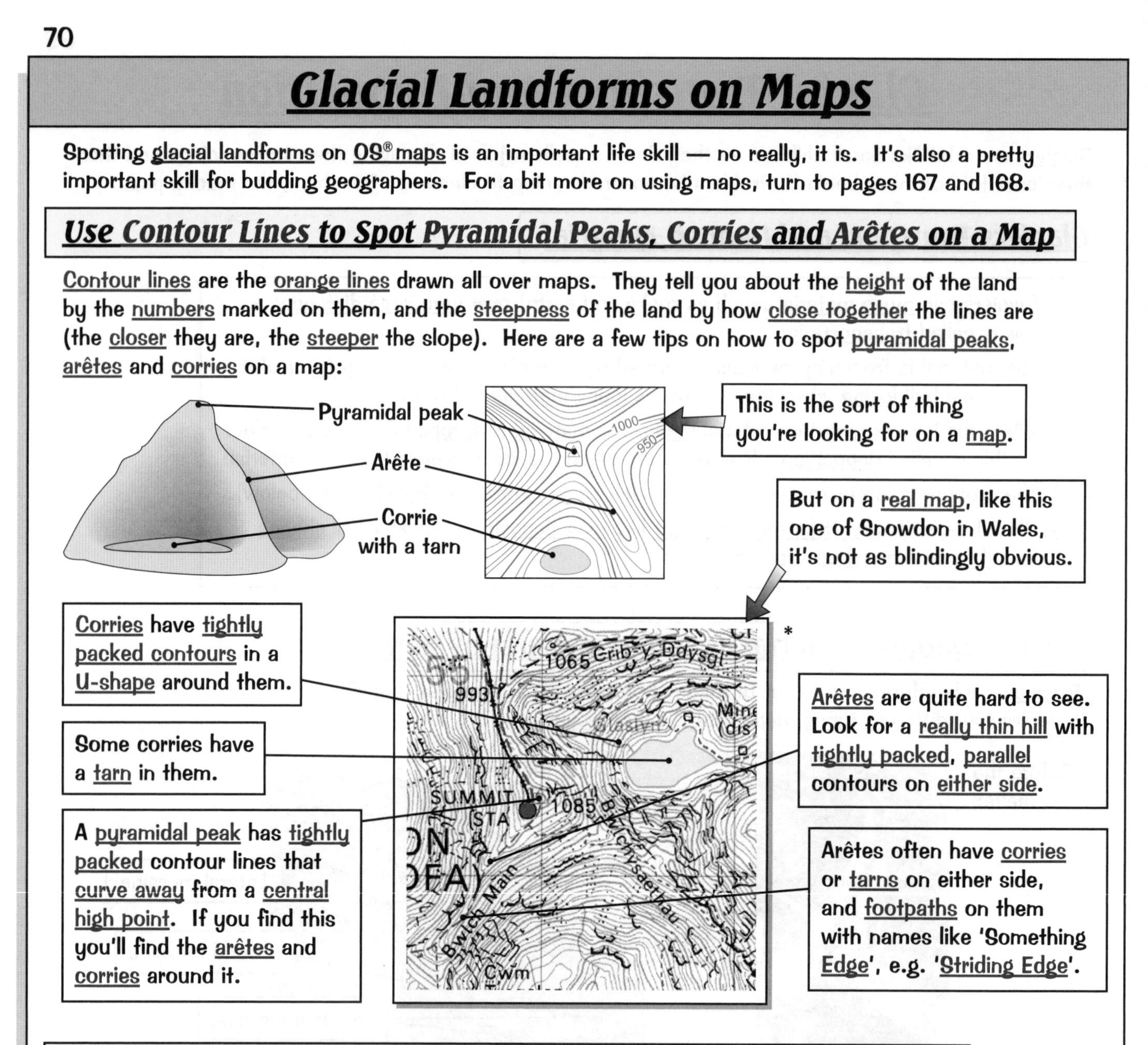

Corries have tightly packed contours in a U-shape around them.

Some corries have a tarn in them.

A pyramidal peak has tightly packed contour lines that curve away from a central high point. If you find this you'll find the arêtes and corries around it.

Arêtes are quite hard to see. Look for a really thin hill with tightly packed, parallel contours on either side.

Arêtes often have corries or tarns on either side, and footpaths on them with names like 'Something Edge', e.g. 'Striding Edge'.

You can also use Maps to Spot Glacial Troughs and Ribbon Lakes

This map of Nant Ffrancon (a glacial trough) and Llyn Ogwen (a ribbon lake) in Wales shows the classic things to look out for if you're ever asked to spot a glacial trough or a ribbon lake on a map extract:

Glacial troughs are flat valleys with very steep sides. There are no contour lines on the bottom of the valley but they're tightly packed on the sides.

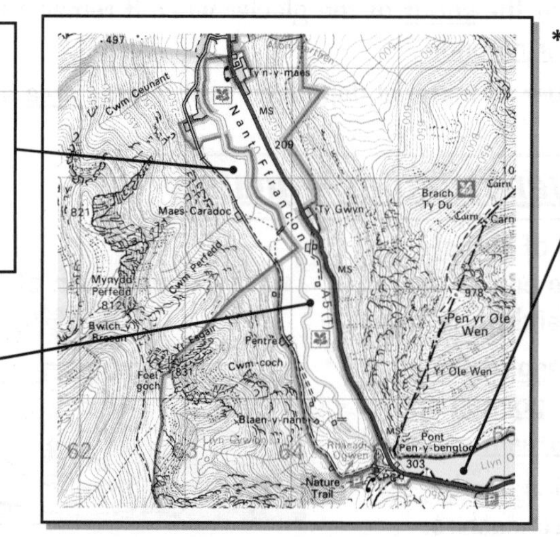

Look for a wide, straight valley in a mountainous area with a river that looks too small to have formed the valley.

Many glacial troughs have ribbon lakes in them. Look for a flat valley with steep sides surrounding a long straight lake.

Con-tours — when the Prison Service goes sight-seeing...

Answering a question using a map shouldn't be too tricky, just study the map carefully and say what you see. Make sure you refer to the map in your answer though — for help with this, e.g. using grid references, see p. 167.

Impacts and Management of Tourism On Ice

We may think skiing holidays are all about lazy days on the slopes and après-ski indulgence, but they have other impacts apart from shrinking our bank balances, and these impacts need to be managed.

Areas Covered in Snow and Ice are Fragile Environments

Areas that are covered in snow and ice attract lots of tourists for things like winter sports and sightseeing of glaciers. The environments are fragile though — they're easily damaged and difficult to manage:

- There's only a short growing season (when there's enough light and warmth for plants to grow) — so plants don't have much time to recover if they're damaged.
- Decay is slow because it's so cold. This means any pollution or litter remains in the environment for a long time.

Tourism has Economic, Social and Environmental Impacts

Tourists have economic, social and environmental impacts in areas covered in snow and ice, so there's conflict over how these areas should be used.
Here are a few of the impacts:

ECONOMIC IMPACTS

1) Lots of new businesses are set up for the tourists, e.g. restaurants, hotels and guiding companies for the sports activities. This boosts the local economy.
2) New businesses means there are job opportunities for people in these remote areas.

SOCIAL IMPACTS

1) Increased numbers of people and businesses mean the infrastructure (roads and railways) becomes congested. This makes it more difficult for local people and tourists to get around.
2) More job opportunities mean that more young people will stay in the area instead of leaving to find work in cities.
3) Tourists can trigger avalanches on ski slopes which can cause injuries and deaths.

ENVIRONMENTAL IMPACTS

1) The fragile glacial environment is damaged by people trampling on the snow and the soil beneath, which causes soil erosion.
2) Glacial landforms like moraines are eroded by people walking on them.
3) There's increased noise, pollution and litter from all the people and traffic in the area.
4) The developments in the area, e.g. buildings and ski lifts, have a visual impact on the environment.

There are Management Strategies to Manage the Different Impacts

There's a need to conserve the fragile environment, but people also have the right to see and experience it. There are different strategies to manage the environment so the impacts of tourism are reduced:

1) Tourists are kept informed of avalanche risks, so they know which areas to avoid. Resorts can build structures to slow and divert the moving snow, plant trees to act as barriers, and set off controlled avalanches to dislodge snow before tourists arrive on the slopes in the morning.
2) Improvements to public transport systems can reduce the amount of traffic and so reduce damage to the environment from pollution.
3) Areas can be set aside as nature reserves. Tourist activity in these areas is limited, so their environmental impact is reduced.

Av-a-lanche break — what Cockney skiers do at midday...

Lots more impacts to get through here. Repeat after me... economic, social and environmental. At least there aren't any political ones for you to remember. Once you know the impacts, check you know how they're managed.

Tourism On Ice — Case Study

So, where to go on holiday... Well, if you're allergic to sea water and know that getting sand in your socks leads to a world of pain, I have a tip for you — rumour has it that Chamonix is nice this time of year.

People go to Chamonix for Winter Sports and Sightseeing

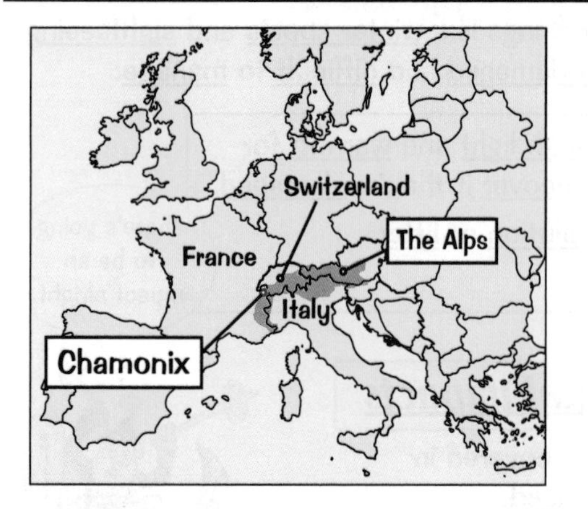

1) The Chamonix Valley is in eastern France at the foot of Mont Blanc (the highest mountain in the Alps). It's close to the border with Italy and Switzerland.
2) It's one of the most popular tourist destinations in the world with around 5 million visitors a year.
3) The region has lots of glaciers, including the Mer de Glace. The Mer de Glace is the longest glacier in France — it's 7 km long and 200 m deep.
4) There are also many other tourist attractions such as 6 ski areas, 350 km of hiking trails, 40 km of mountain bike tracks, an Alpine museum and an exhibition centre.

Tourism has Economic, Social and Environmental Impacts on the Region

ECONOMIC IMPACTS

1) The tourism industry in Chamonix creates a lot of jobs, e.g. 2500 people work as seasonal workers every year.
2) Companies make a lot of money from tourism in Chamonix, e.g. Compagnie du Mont Blanc is a company that runs ski lifts and rail transport — it has a turnover of €50 million.

SOCIAL IMPACTS

1) The types of jobs available in Chamonix have changed from farm labouring to jobs in restaurants and hotels etc.
2) Tourist developments, e.g. ski slopes, have increased the risk of avalanches. This means there are more deaths from avalanches, e.g. in 1999 an avalanche killed 12 people.

ENVIRONMENTAL IMPACTS

1) Large numbers of tourists cause a lot of traffic, which increases pollution. E.g. a study from 2002 to 2004 showed that traffic pollution was worse in the Chamonix region than in the centre of Paris.
2) A huge amount of energy is used to run the facilities for tourists, e.g. the hotels, ski lifts and snow-making machines. This increases CO_2 emissions, which increases global warming.

Tourism in the Resort has to be Carefully Managed

Management of the Chamonix Valley has to balance the need to conserve the environment with the right of people to see and experience it. Here are a few of the strategies used:

1) A system of avalanche barriers is maintained around the resorts, e.g. there's a barrier at Taconnaz. There are also avalanche awareness courses and daily bulletins to keep tourists aware of the risks. This means tourists are less likely to be hurt or killed by an avalanche.
2) The amount of traffic in Chamonix is managed by providing free public transport for tourists. The amount of pollution from public transport is reduced by using low emission buses.
3) Some hotels are reducing their energy use, e.g. by installing solar panels to heat water and systems to automatically turn lights off. This means CO_2 emissions are reduced.

Tourism on Ice — sounds like another piece of TV gold...

Another case study, another day — but at least this one makes you think of holidays. Maybe not the most exciting thing to learn but examiners love to read about 'real world' examples — so the more detail you can shove in the better.

Impacts of Glacial Retreat

The end of this section is near, but don't slack off yet — there's still one page left. Once it's done though, you'll be an expert on all things glacial. So, wave goodbye with a page on the impacts of glacial retreat.

Glacial Retreat and Unreliable Snowfall Affects Tourism

The economies of many areas that are covered in snow and ice rely on money from tourism (e.g. for winter sports and sightseeing of glaciers). These areas are being affected by glacial retreat (see p. 66) and unreliable snowfall:

1) Glacial retreat means the ice will no longer be available for winter sports, e.g. trekking and ice climbing, or sightseeing of glaciers. This means the area will attract fewer tourists.
2) Unreliable snowfall means that there might not be enough snow for winter sports, e.g. skiing and snowboarding. This also means the area will attract fewer tourists.
3) Fewer tourists will mean that the businesses that rely on tourism, e.g. hotels, restaurants and guiding companies, will make less money and may go out of business.
4) This would lead to increased unemployment in these areas.

Retreat!

Retreat! Retreat!

Glacial Retreat has other Impacts

ECONOMIC IMPACTS

Once a glacier has completely melted, the amount of meltwater decreases. This means industries that rely on the supply of meltwater, e.g. agriculture for irrigation and hydroelectric power (HEP) for electricity production, will make less money and could shut down.

SOCIAL IMPACTS

1) Glacial retreat will mean the water supply to some settlements is reduced (see above).
2) Disruptions to power supplies from HEP could leave some people with an unreliable power supply.
3) If businesses shut down, local people will have to move away to find work. Young people in particular will move away, so older family members might be left behind.
4) If an area's population declines, local services and recreational facilities will also shut down.
5) The ice will no longer be available for recreational use for local people, e.g. for trekking and ice climbing.

ENVIRONMENTAL IMPACTS

1) Glacial retreat is linked to an increase in natural hazards — rapid melting can cause flooding, rockslides and avalanches. These hazards destroy habitats and disrupt food chains.
2) Meltwater from retreating glaciers contributes to rising sea level — water is no longer stored as ice on land and returns to the sea. Rising sea level destroys coastal habitats by causing flooding and erosion.
3) Lots of fish species are adapted to live in the cold meltwater that comes from glaciers. When glaciers have completely melted, there's no cold meltwater so these fish species may die out.
4) Harmful pollutants can be trapped in glacial ice, e.g. the pesticide DDT that was used from the 1940s to 1980s. Rapid melting releases them back into the environment, polluting streams and lakes.

Glacial retreat — don't blame the killer whales, they make the grey seals retreat...

Yep, you have now seen off the last chilly, ice-related revision page. When you've finished your celebration of choice slide over to the next page and check it's all firmly stuck in your head. Then you can move on to warmer sections.

Revision Summary for Section 6

Time to whip off your hat and scarf and warm up with some revision questions. The bad news is that if you don't know the answers it would be a good idea to dip back into the icy depths of this section. Once you're confident of all the answers I recommend that you indulge in some kind of recreational activity. Don't disappear for too long though, there's plenty more revision where this section came from.

1) What's the name of the last ice age?
2) How long ago did the last glacial period end?
3) How much of the Earth's land surface is currently covered by ice?
4) What three types of evidence are used to identify past temperature changes?
5) Define the term accumulation.
6) What is the zone of ablation?
7) What happens when a glacier has a positive glacial budget?
8) Why does a glacier retreat in summer?
9) a) Give an example of a retreating glacier.
   b) What evidence is there that the glacier you named has retreated since the 19th century?
   c) Explain why this has happened.
10) What is rotational slip?
11) Explain what freeze-thaw weathering is.
12) What is a corrie?
13) How does a pyramidal peak form?
14) Give an example of a pyramidal peak.
15) Explain how a hanging valley forms.
16) What is bulldozing?
17) Give one difference between lateral and ground moraine.
18) Where is medial moraine deposited?
19) Describe what a drumlin looks like.
20) How would you identify a pyramidal peak on a map?
21) Describe what a glacial trough looks like on a map.
22) Give an example of a glacial trough.
23) What does a ribbon lake look like on a map?
24) Give one reason why areas covered in snow and ice are fragile environments.
25) Give two social impacts of tourism on areas covered in snow and ice.
26) Describe two environmental impacts of tourism on areas covered in snow and ice.
27) a) Give an example of an area in the Alps used for winter sports and sightseeing of glaciers.
   b) Give one economic, one social and one environmental impact of tourism in the area you named.
   c) Give three management strategies used in the area you named.
28) Name an industry affected by glacial retreat.
29) Give one economic impact of glacial retreat.
30) Give one social impact of glacial retreat.

Coastal Weathering and Erosion

Weathering is the breakdown of rocks where they are, erosion is when the rocks are broken down and carried away by something, e.g. by seawater. Poor coastal zone, I bet it's worn down.

Rock is Broken Down by Mechanical and Chemical Weathering

1) Mechanical weathering is the breakdown of rock without changing its chemical composition. There's one main type of mechanical weathering that affects coasts — freeze-thaw weathering:

    1) It happens when the temperature alternates above and below 0 °C (the freezing point of water).
    2) Water gets into rock that has cracks, e.g. granite.
    3) When the water freezes it expands, which puts pressure on the rock.
    4) When the water thaws it contracts, which releases the pressure on the rock.
    5) Repeated freezing and thawing widens the cracks and causes the rock to break up.

2) Chemical weathering is the breakdown of rock by changing its chemical composition. Carbonation weathering is a type of chemical weathering that happens in warm and wet conditions:

    1) Rainwater has carbon dioxide dissolved in it, which makes it a weak carbonic acid.
    2) Carbonic acid reacts with rock that contains calcium carbonate, e.g. carboniferous limestone, so the rocks are dissolved by the rainwater.

Waves Wear Away the Coast using Four Processes of Erosion

1) Hydraulic power — waves crash against rock and compress the air in the cracks. This puts pressure on the rock. Repeated compression widens the cracks and makes bits of rock break off.
2) Abrasion (corrasion) — eroded particles in the water scrape against rock, removing small pieces.
3) Attrition — eroded particles in the water smash into each other and break into smaller fragments. Their edges also get rounded off as they rub together.
4) Solution (corrosion) — weak carbonic acid in seawater dissolves rock like chalk and limestone.

Destructive Waves Erode the Coastline

Coastlines that are being eroded by destructive waves are called destructive coastlines.

The waves that carry out erosional processes are called destructive waves:

1) Destructive waves have a high frequency (10-14 waves per minute).
2) They're high and steep.
3) Their backwash (the movement of the water back down the beach) is more powerful than their swash (the movement of the water up the beach). This means material is removed from the coast.
4) There are two main factors that affect the size and power of destructive waves, and so how much they erode the coast:

    - Wind — the force of the wind on the water's surface is what creates waves. A strong wind gives large, powerful waves.
    - Fetch — is the distance of water over which the wind has blown to produce a wave. The greater the fetch, the bigger and more powerful the wave.

High, steep wave

Backwash

Swash

If you feel like your brain is being eroded — have a little break from revision...

This page is packed full of information, but it's really only about how the coast is worn away and rocks are broken down into smaller pieces. Break your revision down into smaller pieces by learning the processes one at a time.

Coastal Landforms Caused by Erosion

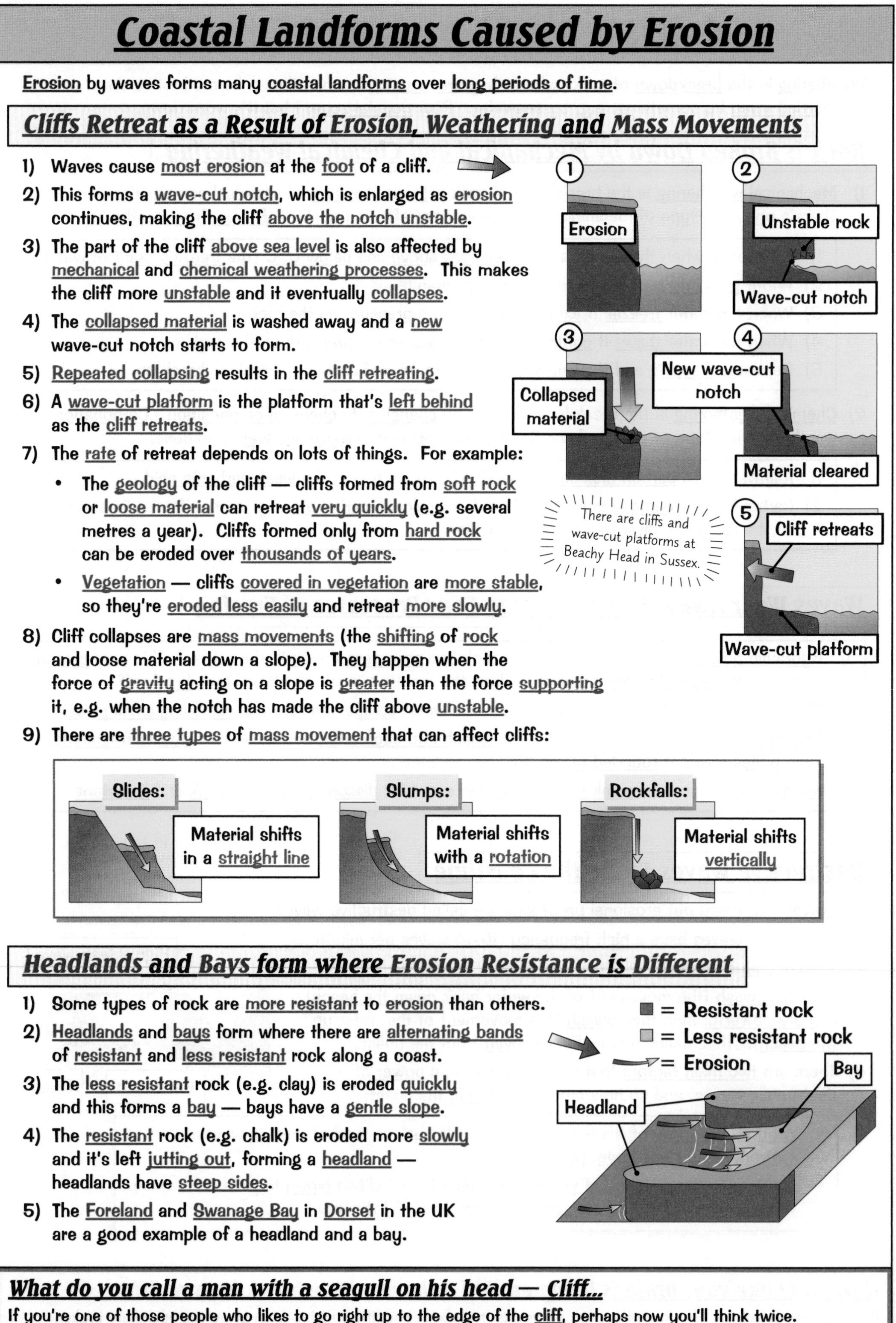

Erosion by waves forms many coastal landforms over long periods of time.

Cliffs Retreat as a Result of Erosion, Weathering and Mass Movements

1) Waves cause most erosion at the foot of a cliff.
2) This forms a wave-cut notch, which is enlarged as erosion continues, making the cliff above the notch unstable.
3) The part of the cliff above sea level is also affected by mechanical and chemical weathering processes. This makes the cliff more unstable and it eventually collapses.
4) The collapsed material is washed away and a new wave-cut notch starts to form.
5) Repeated collapsing results in the cliff retreating.
6) A wave-cut platform is the platform that's left behind as the cliff retreats.
7) The rate of retreat depends on lots of things. For example:
   - The geology of the cliff — cliffs formed from soft rock or loose material can retreat very quickly (e.g. several metres a year). Cliffs formed only from hard rock can be eroded over thousands of years.
   - Vegetation — cliffs covered in vegetation are more stable, so they're eroded less easily and retreat more slowly.
8) Cliff collapses are mass movements (the shifting of rock and loose material down a slope). They happen when the force of gravity acting on a slope is greater than the force supporting it, e.g. when the notch has made the cliff above unstable.
9) There are three types of mass movement that can affect cliffs:

Headlands and Bays form where Erosion Resistance is Different

1) Some types of rock are more resistant to erosion than others.
2) Headlands and bays form where there are alternating bands of resistant and less resistant rock along a coast.
3) The less resistant rock (e.g. clay) is eroded quickly and this forms a bay — bays have a gentle slope.
4) The resistant rock (e.g. chalk) is eroded more slowly and it's left jutting out, forming a headland — headlands have steep sides.
5) The Foreland and Swanage Bay in Dorset in the UK are a good example of a headland and a bay.

What do you call a man with a seagull on his head — Cliff...

If you're one of those people who likes to go right up to the edge of the cliff, perhaps now you'll think twice. There could be an unstable wave-cut notch just below you... not a good way to end your geography career.

Coastal Landforms Caused by Erosion

You're not quite done with coastal erosion yet — it's got a few more tricks up its sleeve...

Headlands are Eroded to form Caves, Arches, Stacks and Stumps

1) Headlands are usually made of resistant rocks (see previous page) that have weaknesses like cracks.
2) Waves crash into the headlands and enlarge the cracks — mainly by hydraulic power and abrasion.
3) Repeated erosion and enlargement of the cracks causes a cave to form.
4) Continued erosion deepens the cave until it breaks through the headland — forming an arch, e.g. Durdle Door in Dorset.
5) Erosion continues to wear away the rock supporting the arch, until it eventually collapses.
6) This forms a stack — an isolated rock that's separate from the headland, e.g. Old Harry in Dorset.
7) The stack is eventually worn away to give a stump, which can be covered by the water at high tide, e.g. Old Harry's Wife in Dorset.

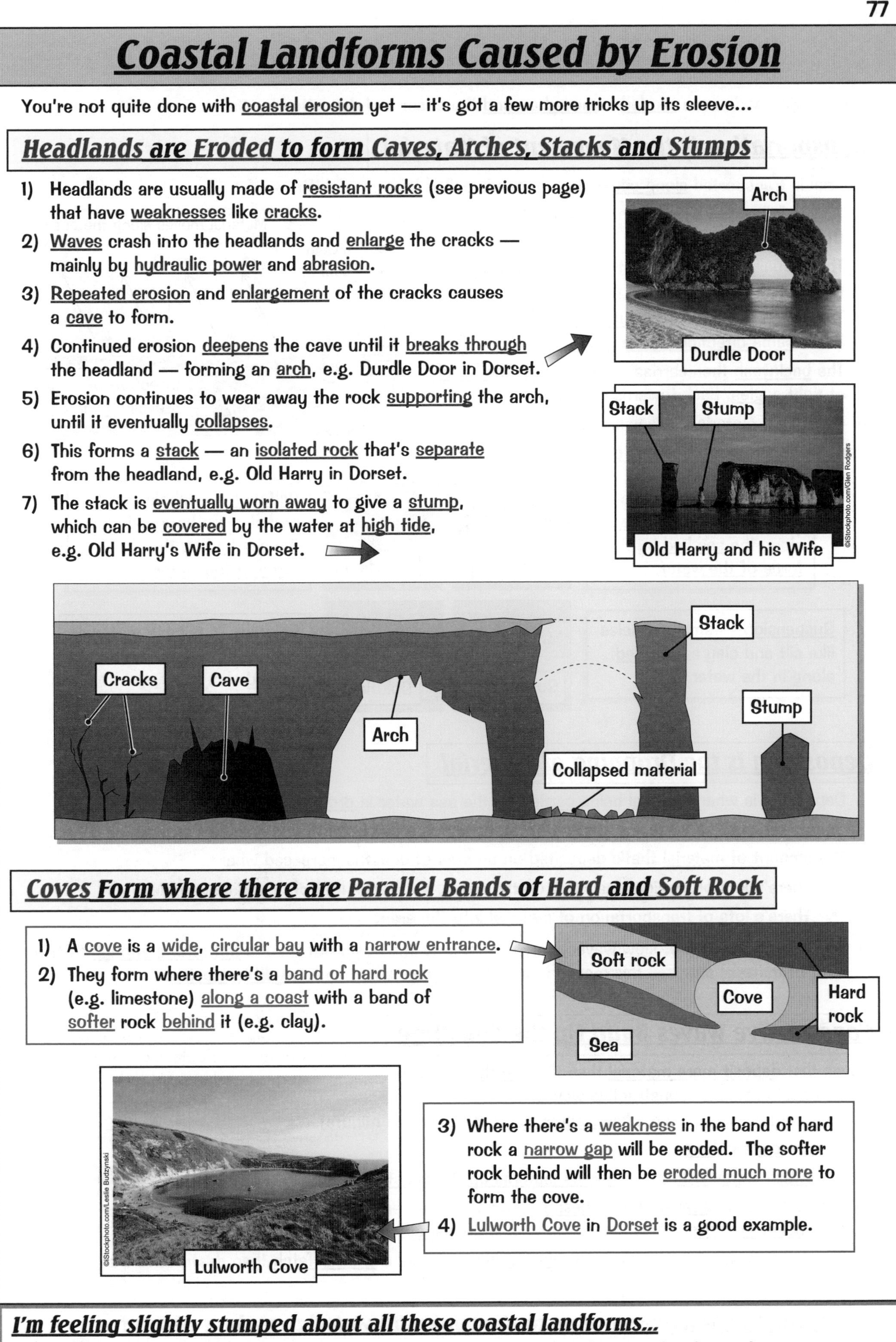

Coves Form where there are Parallel Bands of Hard and Soft Rock

1) A cove is a wide, circular bay with a narrow entrance.
2) They form where there's a band of hard rock (e.g. limestone) along a coast with a band of softer rock behind it (e.g. clay).
3) Where there's a weakness in the band of hard rock a narrow gap will be eroded. The softer rock behind will then be eroded much more to form the cove.
4) Lulworth Cove in Dorset is a good example.

I'm feeling slightly stumped about all these coastal landforms...

Crack erosion — sounds painful. If you don't want it to happen to you, then don't stand in the sea for too many years in a row. Just a few landforms for you here, and as ever, learning the diagrams will help in the exam.

Coastal Transportation and Deposition

The material that's been eroded is moved around the coast and deposited by waves.

Transportation is the Movement of Material

Material is transported along coasts by a process called longshore drift:

1) Waves follow the direction of the prevailing (most common) wind.
2) They usually hit the coast at an oblique angle (any angle that isn't a right angle).
3) The swash carries material up the beach, in the same direction as the waves.
4) The backwash then carries material down the beach at right angles, back towards the sea.
5) Over time, material zigzags along the coast.

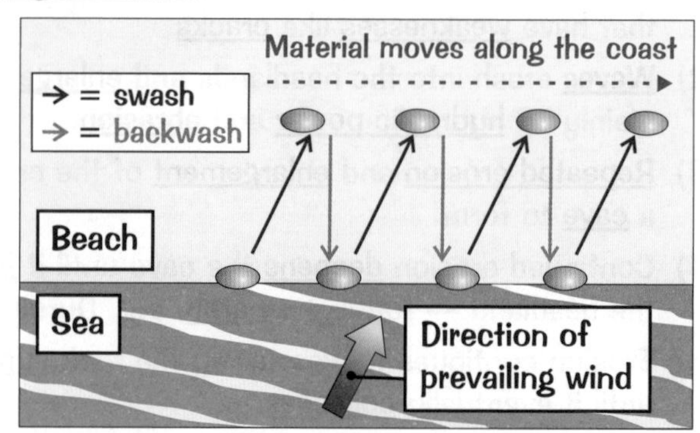

There are four other processes of transportation:

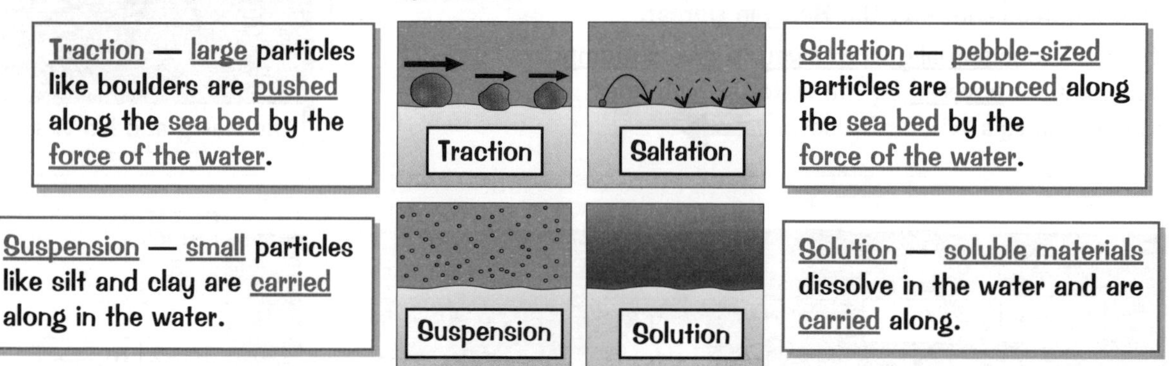

Deposition is the Dropping of Material

1) Deposition is when material being carried by the sea water is dropped on the coast.
2) Coasts are built up when the amount of deposition is greater than the amount of erosion.
3) The amount of material that's deposited on an area of coast is increased when:
   - There's lots of erosion elsewhere on the coast, so there's lots of material available.
   - There's lots of transportation of material into the area.
4) Low energy waves (i.e. slow waves) carry material to the coast but they're not strong enough to take a lot of material away — this means there's lots of deposition and very little erosion.

Constructive Waves Build Up the Coastline

Coastlines being built up by constructive waves are called constructive coastlines.

Waves that deposit more material than they erode and build up the coast are called constructive waves.

1) Constructive waves have a low frequency (6-8 waves per minute).
2) They're low and long.
3) The swash is powerful and it carries material up the coast.
4) The backwash is weaker and it doesn't take a lot of material back down the coast. This means material is deposited on the coast.
5) Constructive waves are made by weaker winds and have a shorter fetch than destructive waves.

Low, long wave

Backwash

Swash

Why did the constructive wave go to the bank? It wanted to make a deposit...

Some more processes for you here but none of them are tricky. You might find it useful to draw yourself a diagram of how longshore drift works — you'll get a feel for how the material is moved along the coast in a zigzag pattern.

Coastal Landforms Caused by Deposition

Here are some more exciting landforms for you to read about. This time it's all about deposition. Unfortunately you're going to be slightly disappointed — sandcastles won't be in the exam.

Beaches are formed by Deposition

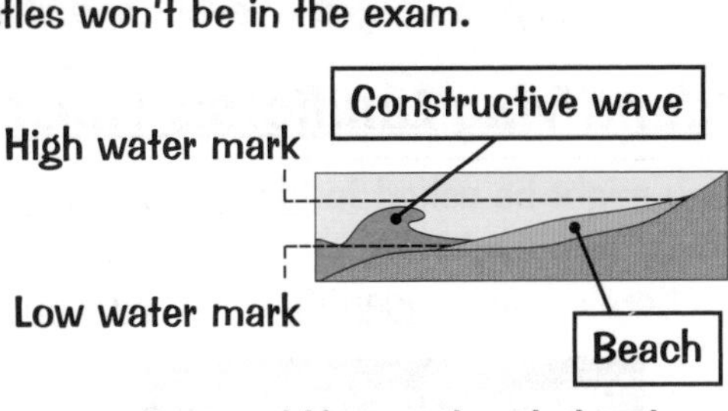

1) Beaches are found on coasts between the high water mark (the highest point on the land the sea level gets to) and the low water mark (the lowest point on the land the sea level gets to).
2) They're formed by constructive waves (see previous page) depositing material like sand and shingle.
3) Sand and shingle beaches have different characteristics:

- Sand beaches are flat and wide — sand particles are small and the weak backwash can move them back down the beach, creating a long, gentle slope.
- Shingle beaches are steep and narrow — shingle particles are large and the weak backwash can't move them back down the beach. The shingle particles build up and create a steep slope.

Spits and Bars are formed by Longshore Drift

Spits are just beaches that stick out into the sea — they're joined to the coast at one end. If a spit sticks out so far that it connects with another bit of the mainland, it'll form a bar. Spits and bars are formed by the process of longshore drift (see previous page).

SPITS

1) Spits form at sharp bends in the coastline, e.g. at a river mouth.
2) Longshore drift transports sand and shingle past the bend and deposits it in the sea.
3) Strong winds and waves can curve the end of the spit (forming a recurved end).
4) The sheltered area behind the spit is protected from waves — lots of material accumulates in this area, which means plants can grow there.
5) Over time, the sheltered area can become a mud flat or a salt marsh.

An example of a spit is Spurn Head in Yorkshire.

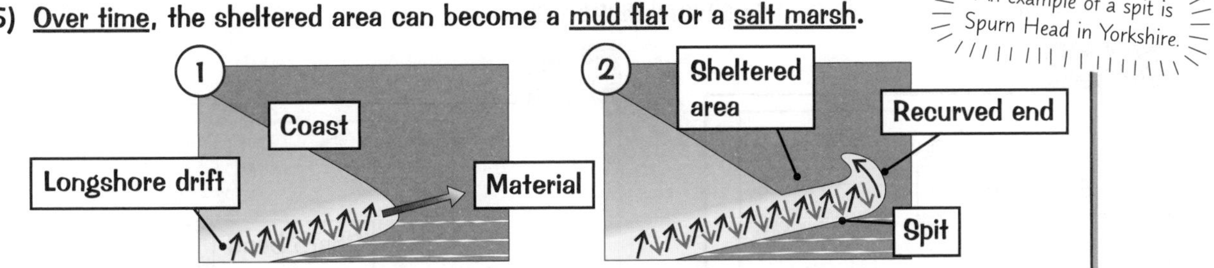

BARS

1) A bar is formed when a spit joins two headlands together, e.g. there's a bar at Slapton in Devon.
2) The bar cuts off the bay between the headlands from the sea.
3) This means a lagoon can form behind the bar.
4) A bar that connects the shore to an island (often a stack) is called a tombolo. For example, Chesil Beach in Dorset joins to the Isle of Portland.

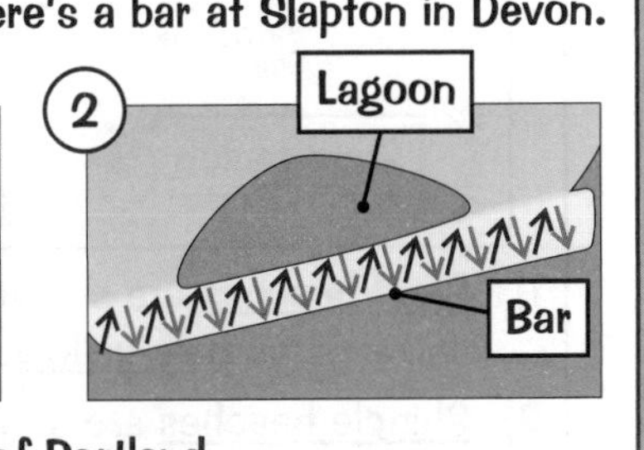

Depositional bars — the only cocktail they serve is a long beach iced tea...

The things you learn in geography are life skills. If you have to meet someone at the beach you now know exactly where to stand — between the high and low water marks, of course. And who said geography was just about maps...

Coastal Landforms on Maps

I love **maps**, all geographers love maps. I can't get to sleep unless I've got one under my pillow. So I'm going to do you a favour and share my passion with you — check out these **coastal landforms**...

Identifying Landforms Caused by Erosion

Have a gander at pages 167-168 for more on reading maps.

You might be asked to **identify coastal landforms** on a **map** in the exam. The simplest thing they could ask is whether the map is showing **erosional** or **depositional landforms**, so here's how to **identify** a few **erosional landforms** to get you started:

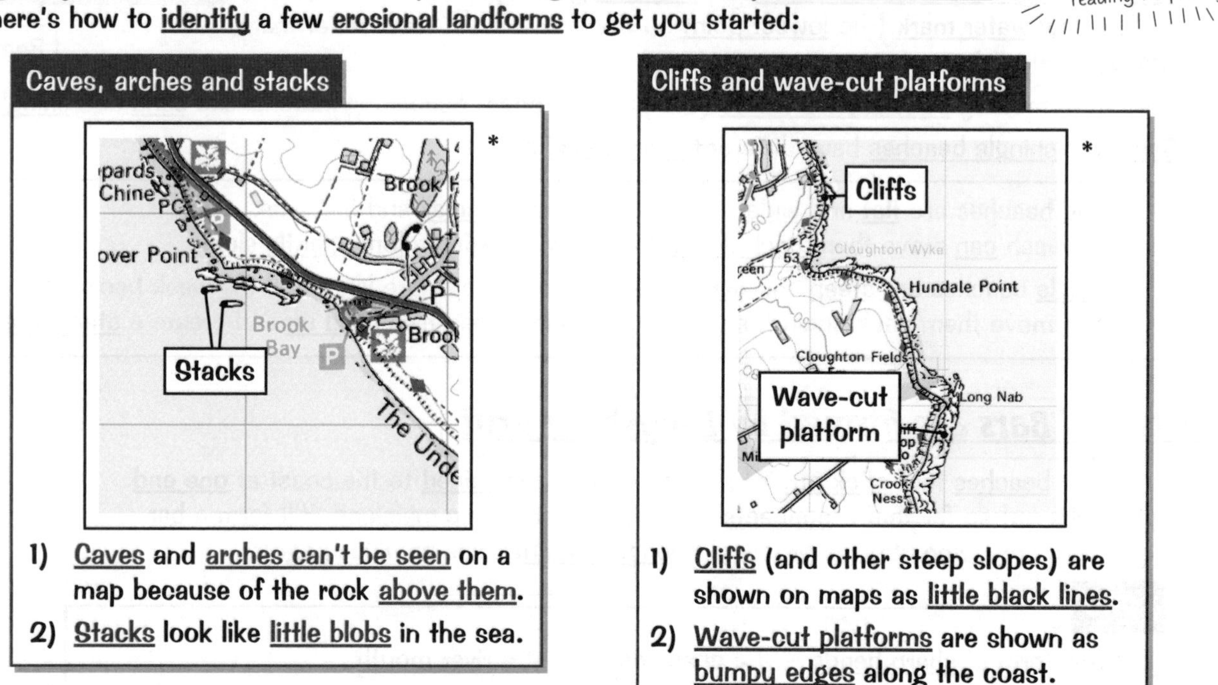

Caves, arches and stacks

1) **Caves** and **arches can't be seen** on a map because of the rock **above them**.
2) **Stacks** look like **little blobs** in the sea.

Cliffs and wave-cut platforms

1) **Cliffs** (and other steep slopes) are shown on maps as **little black lines**.
2) **Wave-cut platforms** are shown as **bumpy edges** along the coast.

Identifying Landforms Caused by Deposition

Identifying depositional landforms is easy once you know that **beaches** are shown in **yellow** on maps. Here's how to **identify** a couple of **depositional landforms**:

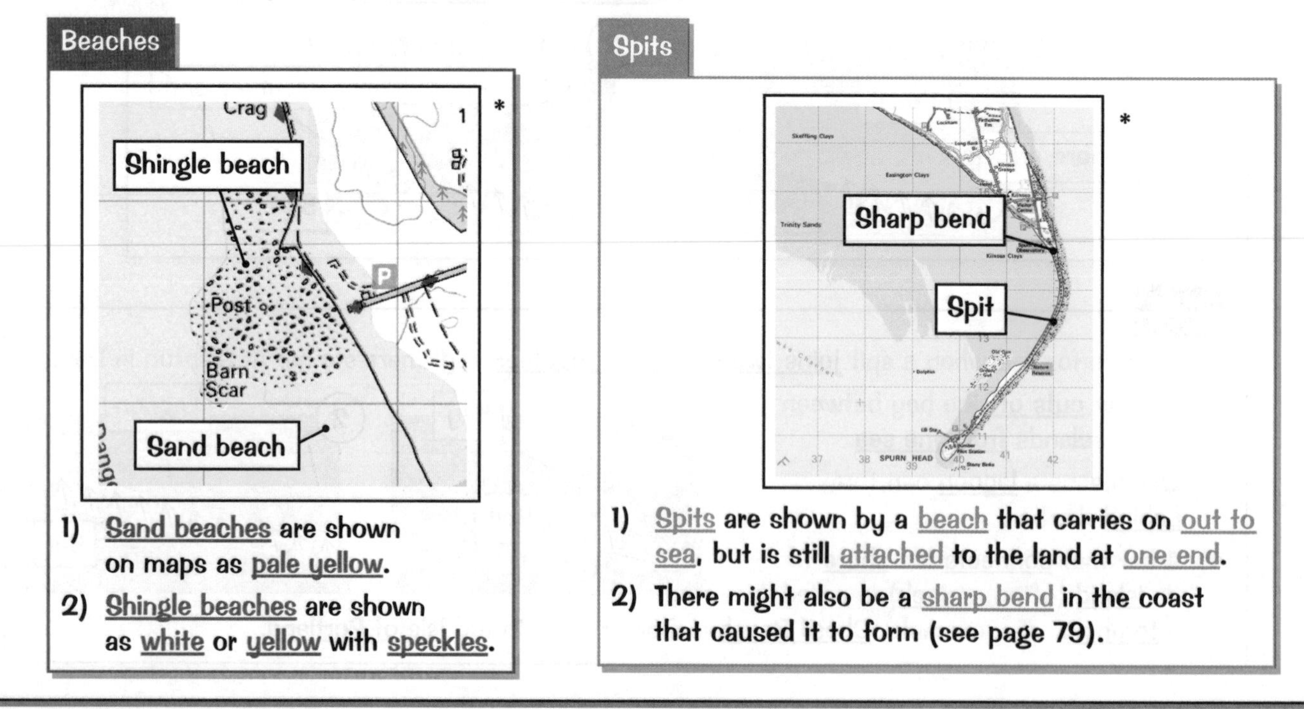

Beaches

1) **Sand beaches** are shown on maps as **pale yellow**.
2) **Shingle beaches** are shown as **white** or **yellow** with **speckles**.

Spits

1) **Spits** are shown by a **beach** that carries on **out to sea**, but is still **attached** to the land at **one end**.
2) There might also be a **sharp bend** in the coast that caused it to form (see page 79).

Find the spit on the map — and then wipe it off...

There are some seriously easy marks up for grabs with map questions so this is a really **useful** page. You could practise looking for **landforms** on any **maps** you can get a hold of. Probably best to avoid the inner-city maps though.

Coastal Area — Case Study

If coastal landforms are your thing (and let's face it, how could they not be), then the Dorset coast is paradise on Earth. It's got the lot — headlands, bays, arches, stacks, coves, tombolos, lagoons...

The Dorset Coast has Examples of many Coastal Landforms

The Dorset coast is made from bands of hard rock (like limestone and chalk) and soft rock (like clay). The rocks have been eroded at different rates giving headlands and bays and lots of other exciting coastal landforms.

Durdle Door

Durdle Door is a great example of an arch. Erosion by waves opened up a crack in the limestone headland, which became a cave and then developed into an arch.

Lulworth Cove

Lulworth Cove is a cove formed after a gap was eroded in a band of limestone. Behind the limestone is a band of clay, which has been eroded away to form the cove. The same is now starting to happen at Stair Hole further west along the coast.

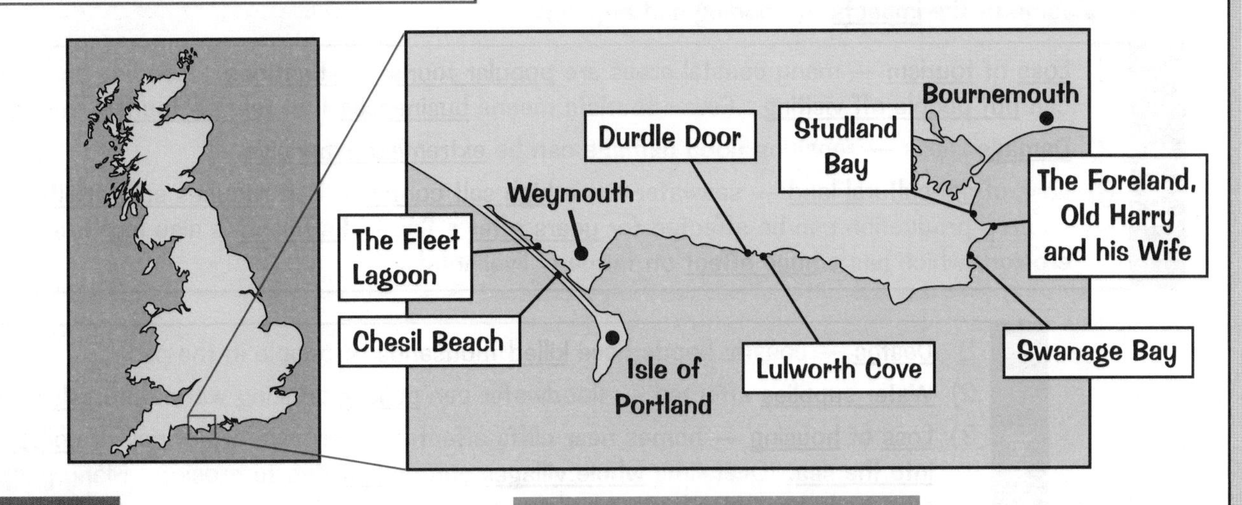

Chesil Beach

Chesil Beach is a tombolo formed by longshore drift. It joins the Isle of Portland to the mainland. Behind Chesil Beach is a shallow lagoon called The Fleet Lagoon.

Swanage Bay and Studland Bay

There are two bays with beaches called Swanage Bay and Studland Bay. They're areas of softer rock (sandstone and clay). In between them is a headland called The Foreland made from a band of harder rock (chalk). The end of the headland has been eroded to become a stack called Old Harry and a stump called Old Harry's Wife.

I love a good tombolo — prizes and fun for everyone...

That's actually Old Harry's second wife. His first wife collapsed into the sea in 1896. It was sad, but she would've wanted him to move on. Before you move on, check that you know the names of the landforms of the Dorset coast.

Rising Sea Level, Coastal Flooding and Erosion

Rising sea level — there's no need to worry if you live in a block of flats... or on top of Mount Everest.

Sea Level is Rising because of Global Warming

Global sea level is rising at a rate of about 2 mm per year. That might not sound like a lot, but sea level has increased by about 20 cm over the past century. It's predicted to rise by between 18 and 59 cm by 2100. The cause of rising sea level is global warming — the rapid rise in global temperature over the last 100 years. Global warming has two effects that cause sea level to rise:

1) Melting ice
The melting of ice on land (e.g. the Antarctic ice sheet) causes water that's stored as ice to return to the oceans. This increases the volume of water in the oceans and causes sea level to rise.

2) Heating oceans
Increased global temperature causes the oceans to get warmer and expand (thermal expansion). This increases the volume of water, causing sea level to rise.

Coastal Areas are at Risk from Flooding and Erosion

Rising sea level will mean coastal flooding will happen more often and will cause more damage, especially in low-lying parts of the world like Bangladesh and the Maldives. Coastal erosion (see page 75) also causes lots of damage along coastlines.
Here are some of the impacts of flooding and erosion:

Economic

1) Loss of tourism — many coastal areas are popular tourist destinations. Flooding and erosion can put people off visiting. Fewer tourists means businesses that rely on tourism may close.
2) Damage repair — repairing flood damage can be extremely expensive.
3) Loss of agricultural land — seawater has a high salt content. Salt reduces soil fertility, so crop production can be affected for years after a flood. Farmland is also lost to coastal erosion, which has a huge effect on farmers' livelihoods.

Social

1) Deaths — coastal floods have killed thousands of people in the past.
2) Water supplies affected — floodwater can pollute drinking water with salt or sewage.
3) Loss of housing — homes near cliffs affected by erosion are at risk of collapsing into the sea. Over time whole villages can be lost due to erosion. Many people are also made homeless because of floods.
4) Loss of jobs — coastal industries may be shut down because of damage to equipment and buildings by floods, e.g. fishing boats can be destroyed.

Political

The government has to make policies to reduce the impacts of future flooding and erosion. They can do things like building more or better coastal defences, or they can manage the use of areas that might be flooded or eroded, e.g. by stopping people living there.

Environmental

1) Ecosystems affected by flooding — seawater has a high salt content. Increased salt levels can damage or kill organisms in an ecosystem. The force of floodwater also uproots trees and plants, and standing flood water also drowns some trees and plants.
2) Loss of habitats — wildlife habitats can be destroyed when coastlines are eroded, e.g. when cliffs collapse (see page 76).

The Antarctic ice sheet is the biggest ice cube in the world...

As you've just discovered, coastal flooding and erosion can have enormous impacts on coastal areas. I know what you're thinking, something's missing. I know what it is — case studies... but don't worry, they're only a page away.

Coastal Flooding — Case Study

Another case to study — this one's all about the impacts of coastal flooding on the Maldives. Enjoy...

The Maldives is a Group of Islands in the Indian Ocean

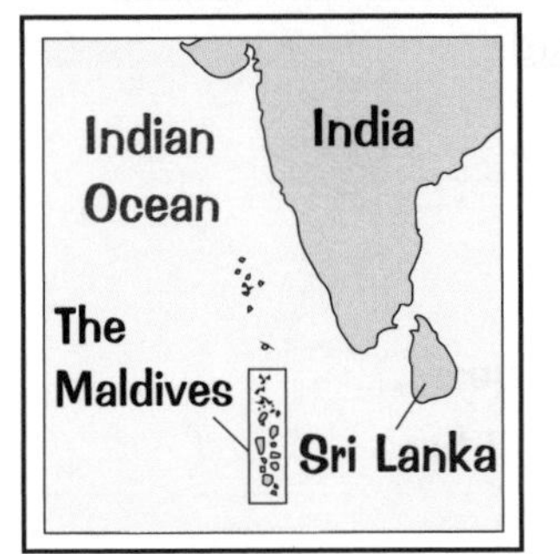

Population: About 300 000 people.

Number of islands: 1190, of which 199 are inhabited.

Average island height: 1.5 m above sea level — 80% of the land is below 1 m. Because of rising sea levels, scientists think the islands will be completely submerged within 50 to 100 years.

Coastal Flooding has a Variety of Impacts on the Maldives

Economic

1) Loss of tourism — tourism is the largest industry in the Maldives. If the main airport can't work properly because of coastal flooding the country will be cut off from international tourists. This will massively reduce the country's income.
2) Disrupted fishing industry — fish are the Maldives' largest export. Coastal flooding may damage fish processing plants, reducing the fish exports and the country's income.

Social

1) Houses damaged or destroyed — a severe flood could make entire communities homeless.
2) Less freshwater available — supplies of freshwater are already low on many of the islands. If supplies are polluted with salty seawater during floods, then some islands will have to rely on rainwater or build expensive desalination plants to meet their water demands.

Environmental

1) Loss of beaches — coastal flooding wears away beaches on the islands at a rapid rate. This destroys habitats and exposes the land behind the beach to the effects of flooding.
2) Loss of soil — the soil on most of the islands is shallow (about 20 cm deep or less). Coastal floods could easily wash away the soil layer, which would mean most plants won't be able to grow.

Political

1) The Maldivian Government had to ask the Japanese Government to give them $60 million to build the 3 m high sea wall that protects the capital city, Malé.
2) Changes to environmental policies — increased flooding is caused by rising sea level, which is caused by global warming (see page 82). The Maldives has pledged to become carbon neutral so it doesn't contribute to global warming. The Maldivian Government is encouraging other governments to do the same.
3) Changes to long-term plans — the government is thinking about buying land in countries like India and Australia and moving Maldivians there, before the islands become uninhabitable.

Carbon neutral means not adding carbon dioxide (CO_2) to the atmosphere — CO_2 is thought to be causing global warming.

It's a lovely place to visit — particularly if you don't like walking up hills...

The Maldives are as flat as a pancake and that's not ideal when you're surrounded by the sea. Now you've read this page you'll know about the impacts that coastal flooding is having on the country. In short, it's not looking good...

Coastal Erosion — Case Study

Holderness in east Yorkshire has one of the fastest eroding coastlines in Europe. What a claim to fame...

The Average Rate of Erosion at Holderness is About 1.8 Metres per Year

1) The Holderness coastline is 61 km long — it stretches from Flamborough Head (a headland) to Spurn Head (a spit).
2) Erosion is causing the cliffs to collapse along the coastline. The material then gets washed away, so the coastline is retreating.
3) About 1.8 m of land is lost to the sea every year — in some places, e.g. Great Cowden, the rate of erosion has been over 10 m per year in recent years.
4) Here are the main reasons for this rapid erosion at Holderness and the impact it has on people's lives and the environment:

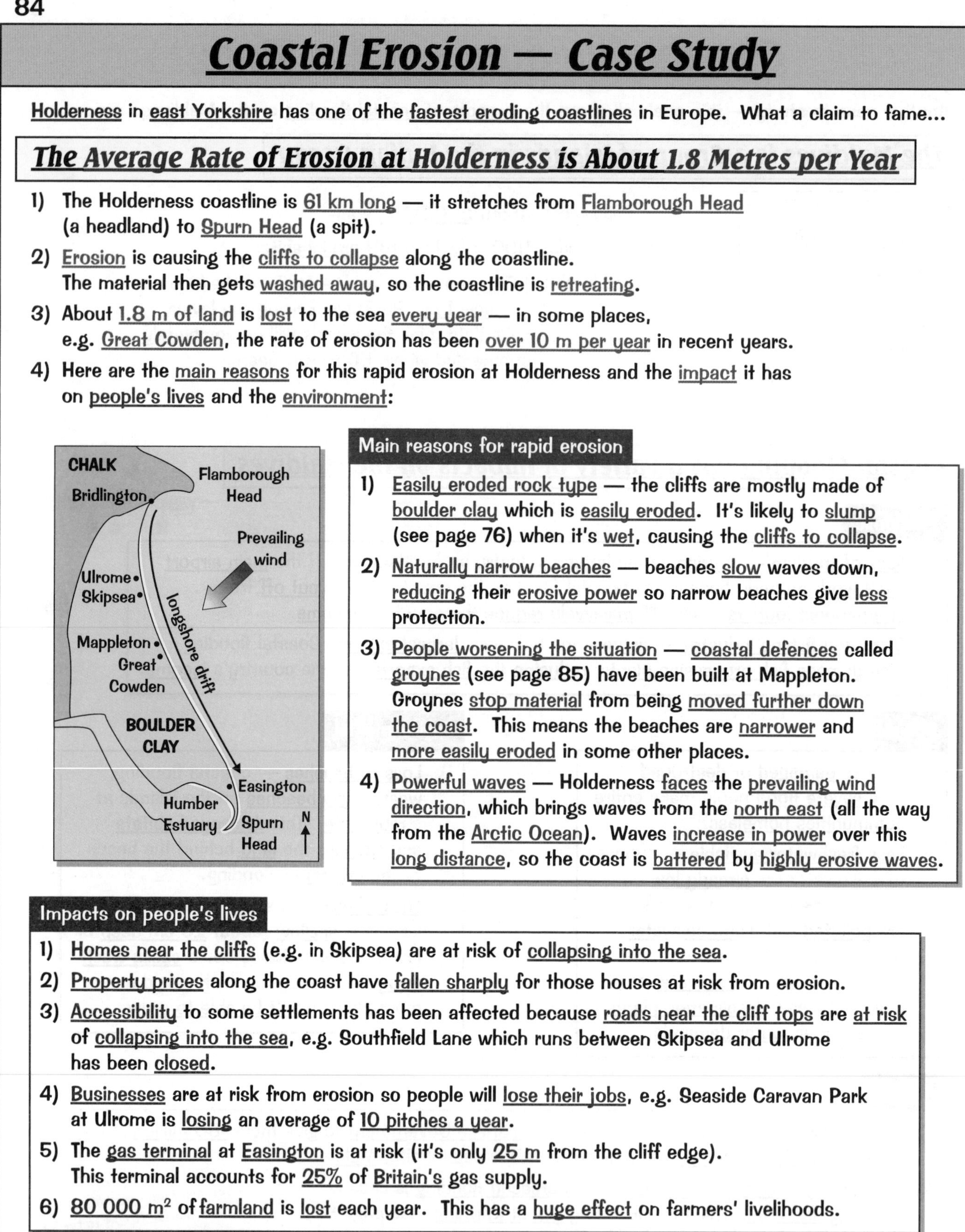

Main reasons for rapid erosion

1) Easily eroded rock type — the cliffs are mostly made of boulder clay which is easily eroded. It's likely to slump (see page 76) when it's wet, causing the cliffs to collapse.
2) Naturally narrow beaches — beaches slow waves down, reducing their erosive power so narrow beaches give less protection.
3) People worsening the situation — coastal defences called groynes (see page 85) have been built at Mappleton. Groynes stop material from being moved further down the coast. This means the beaches are narrower and more easily eroded in some other places.
4) Powerful waves — Holderness faces the prevailing wind direction, which brings waves from the north east (all the way from the Arctic Ocean). Waves increase in power over this long distance, so the coast is battered by highly erosive waves.

Impacts on people's lives

1) Homes near the cliffs (e.g. in Skipsea) are at risk of collapsing into the sea.
2) Property prices along the coast have fallen sharply for those houses at risk from erosion.
3) Accessibility to some settlements has been affected because roads near the cliff tops are at risk of collapsing into the sea, e.g. Southfield Lane which runs between Skipsea and Ulrome has been closed.
4) Businesses are at risk from erosion so people will lose their jobs, e.g. Seaside Caravan Park at Ulrome is losing an average of 10 pitches a year.
5) The gas terminal at Easington is at risk (it's only 25 m from the cliff edge). This terminal accounts for 25% of Britain's gas supply.
6) 80 000 m^2 of farmland is lost each year. This has a huge effect on farmers' livelihoods.

Environmental impacts

Some SSSIs (Sites of Special Scientific Interest) are threatened — e.g. the Lagoons near Easington are part of an SSSI. The Lagoons are separated from the sea by a narrow strip of sand and shingle (a bar). If this is eroded it will connect the Lagoons to the sea and they would be destroyed.

Extreme caravanning — the latest craze in Ulrome...

Blimey, Holderness really is taking a battering from the sea. See if you can remember the causes and the impacts of the rapid erosion — cover the page and write down three of each to find out what you know. To the next page...

Coastal Management Strategies

The aim of coastal management is to protect people and the environment from the impacts of erosion and flooding. Unfortunately it's not as simple as a big fence and a bucket, nice though that would be.

Coastal Defences Include Hard and Soft Engineering

There are two types of strategy to deal with coastal flooding and erosion:

Hard engineering — man-made structures built to control the flow of the sea and reduce flooding and erosion.

Soft engineering — schemes set up using knowledge of the sea and its processes to reduce the effects of flooding and erosion.

	Strategy	What it is	Benefits	Disadvantages
HARD ENGINEERING	Sea wall	A wall made out of a hard material like concrete that reflects waves back to sea.	It prevents erosion of the coast. It also acts as a barrier to prevent flooding.	It creates a strong backwash, which erodes under the wall. Sea walls are very expensive to build and to maintain.
HARD ENGINEERING	Rock armour	Boulders that are piled up along the coast.	The boulders absorb wave energy and so reduce erosion and flooding.	Boulders can be moved around by strong waves, so they need to be replaced.
HARD ENGINEERING	Groynes	Wooden or stone fences that are built at right angles to the coast. They trap material transported by longshore drift.	Groynes create wider beaches which slow the waves. This gives greater protection from flooding and erosion.	They starve beaches further down the coast of sand, making them narrower. Narrower beaches don't protect the coast as well, leading to greater erosion and floods.
HARD ENGINEERING	Breakwaters	Concrete blocks or boulders deposited on the sea bed off the coast.	They force waves to break offshore so their erosive power is reduced before they reach the shore.	They're expensive and can be damaged by storms.
SOFT ENGINEERING	Beach nourishment	Sand and shingle from elsewhere (e.g. the offshore seabed) that's added to beaches.	Beach replenishment creates wider beaches which slow the waves. This gives greater protection from flooding and erosion.	Taking material from the seabed can kill organisms like sponges and corals. It's a very expensive defence. It has to be repeated.
SOFT ENGINEERING	Dune regeneration	Creating or restoring sand dunes by either nourishment, or by planting vegetation to stabilise the sand.	Sand dunes provide a barrier between the land and the sea. Wave energy is absorbed which prevents flooding and erosion. Stabilisation is cheap.	The protection is limited to a small area. Nourishment is very expensive.
SOFT ENGINEERING	Marsh creation	Planting vegetation in mudflats along the coast.	The vegetation stabilises the mudflats and helps to reduce the speed of the waves. This prevents flooding and erosion. It also creates new habitats for organisms.	Marsh creation isn't useful where erosion rates are high because the marsh can't establish itself. It's a fairly expensive defence.
SOFT ENGINEERING	Managed retreat	Removing an existing defence and allowing the land behind it to flood.	Over time the land will become marshland — creating new habitats. Flooding and erosion are reduced behind the marshland. It's a fairly cheap defence.	People may disagree over what land is allowed to flood, e.g. flooding farmland would affect the livelihood of farmers.

Some strategies for coastal management are more sustainable than others. Sustainable strategies meet the needs of people today (i.e. they reduce flooding and erosion), without stopping people in the future getting the things they need. This means not using up too many resources (e.g. money) or damaging the environment.

Hard engineering strategies are usually less sustainable than soft engineering strategies because they generally cost more money to build and maintain, and they damage the environment more.

A good management strategy involves warm-ups — they reduce groyne strains...

Wow, that sure is a mighty fine table. It seems like a lot to remember but I promise you, it's really not that tough. Try to hold on to a couple of benefits and disadvantages for each strategy — you'll thank me later, honest.

Coastal Management — Case Study

We're going back to Holderness to find out what management strategies are being used there. You know, it's possible that we're having too much fun...

Hard Engineering Strategies have been used Along Holderness

Page 84 outlines the main reasons for the rapid erosion along Holderness and the impacts it's having. To try to reduce the effects of erosion, 11.4 km of Holderness coastline has been protected by hard engineering:

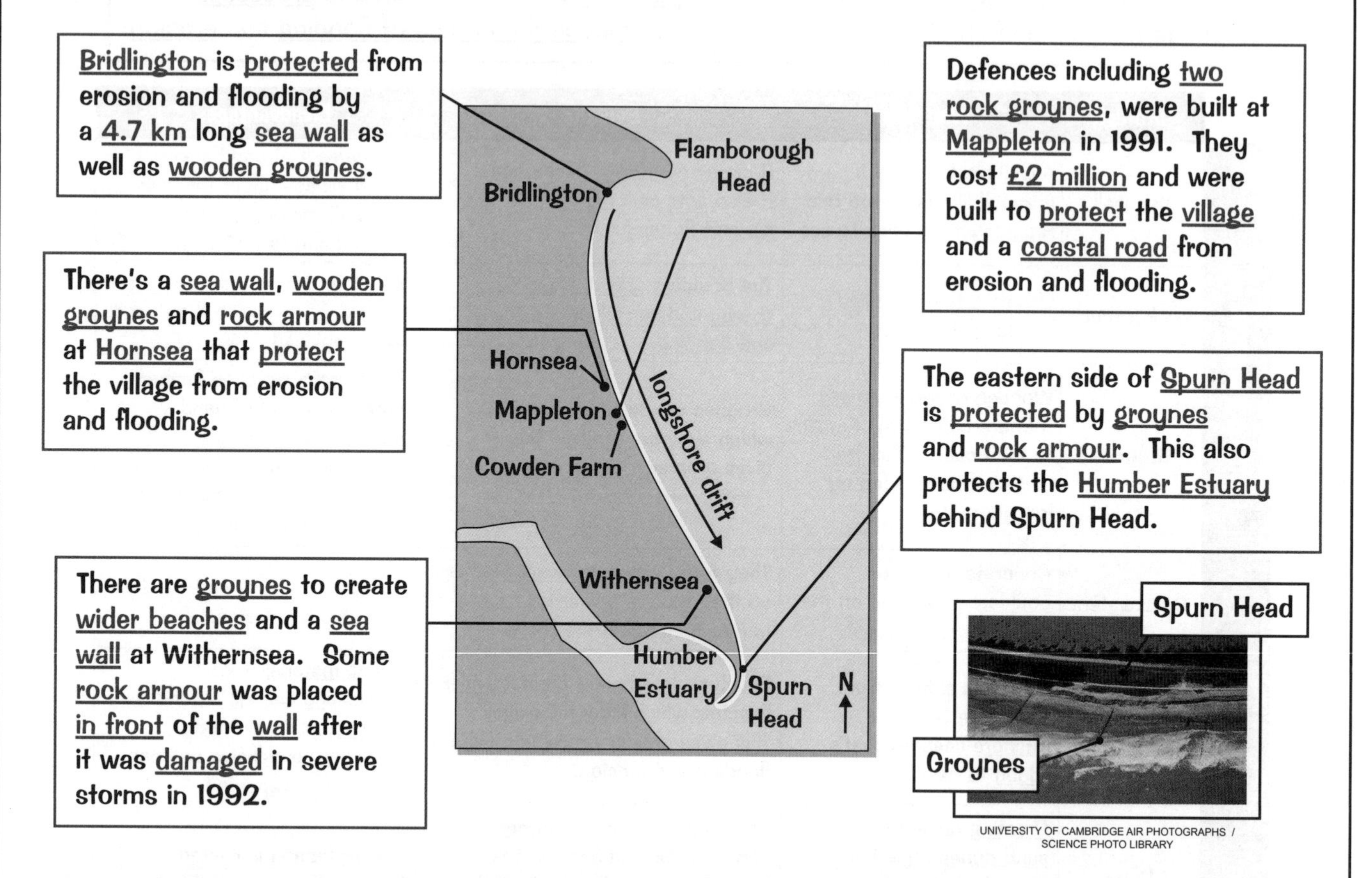

UNIVERSITY OF CAMBRIDGE AIR PHOTOGRAPHS / SCIENCE PHOTO LIBRARY

The Strategies are Locally Successful but Cause Problems Elsewhere

1) Groynes protect local areas but cause narrow beaches to form further down the Holderness coast. This increases erosion down the coast, e.g. Cowden Farm (south of Mappleton) is now at risk of falling into the sea.
2) The material produced from the erosion of Holderness is normally transported south into the Humber Estuary and down the Lincolnshire coast. Reducing the amount of material that's eroded and transported south increases the risk of flooding in the Humber Estuary, because there's less material to slow the floodwater down.
3) The rate of coastal retreat along the Lincolnshire coast is also increased, because less new material is being added.
4) Spurn Head is at risk of being eroded away because less material is being added to it.
5) Bays are forming between the protected areas, and the protected areas are becoming headlands which are being eroded more heavily. This means maintaining the defences in the protected areas is becoming more expensive.

Holderness coastal management officer — probably not the easiest job to have...

This follows on from the case study on page 84, so hopefully most of the place names seem familiar. You'll probably be able to use information from both pages to answer a question on coastal management. What an educational bargain.

Coastal Habitat — Case Study

Coastal areas get pretty heavily used by people (beach bums), but they're important for wildlife too...

Studland Bay is a Coastal Area with Beaches, Dunes and Heathland

1) Studland Bay is a bay in Dorset, in the south west of England.
2) It's mostly sheltered from highly erosive waves, but the southern end of the bay is being eroded.
3) There are sandy beaches around the bay, with sand dunes and heathland behind them.
4) The heathland is a Site of Special Scientific Interest (SSSI) and a nature reserve.
5) Studland Bay is also a popular tourist destination.

Studland Bay Provides a Habitat for a Large Variety of Wildlife

Here are a few examples of the wildlife that's found in Studland Bay:

- Reptiles like adders, grass snakes, sand lizards and slow worms.
- Birds like Dartford warblers (a rare bird in England), shelducks and grebes.
- Fish like seahorses — Studland Bay is the only place in Britain where the spiny seahorse breeds.
- Plants like marram grass and lyme grass on the sand dunes and heather on the heathland.

Some of these organisms are specially adapted to live in the habitats found in Studland Bay:

1) Marram grass has folded leaves to reduce water loss — sand dunes are windy and dry which increases transpiration. It also has long roots to take up water and to stabilise itself in the loose sand.
2) Lyme grass has waxy leaves to reduce water loss by transpiration.
3) Grebes — these birds dive underwater to find food in the sea. Their feet are far back on their bodies to help them dive (it makes them streamlined).
4) Snakes and lizards have thick, scaly skin to reduce water loss from their bodies. It also protects them from rough undergrowth on the heathland.

Transpiration is the loss of water from plants by evaporation.

There are Conflicts Between Land Use and the Need for Conservation

Some human activities (e.g. recreation) don't use the environment in a sustainable way (they use up resources or damage the environment). The environment is managed to make sure it's conserved, but can also be used for other activities:

1) Lots of people walk across the sand dunes which has caused lots of erosion. The National Trust manages the area so people can use the sand dunes without damaging them too much:
   - Boardwalks are used to guide people over the dunes so the sand beneath them is protected.
   - Some sand dunes have been fenced off and marram grass has been planted in them. This gives the dunes a chance to recover and the marram grass stabilises the sand.
   - Information signs have been put up to let visitors know why the sand dune habitat is important, and how they can enjoy the environment without damaging it.
2) Hundreds of boats use Studland Bay and their anchors are destroying the seagrass where seahorses live. Seahorses are protected by law, so boat owners are being told to not damage the seagrass.
3) The heathland behind the sand dunes is an important habitat, but it can be damaged by fires caused by things like cigarettes, e.g. in 2008 a fire destroyed six acres of heathland. The National Trust is educating visitors on the dangers of causing fires and has provided fire beaters to extinguish flames.

Stud-land bay — sorry girls, it's not what you're thinking of...

Plenty of words here and not very many pretty pictures I'm afraid. It'll just be a case of cramming all these facts into your head so you can cough them up on demand. Just think of the seahorses and it'll all be worth it in the end.

Revision Summary for Section 7

So, you've coasted through another section — that means it's time to find out just how much of this information has been deposited in your noggin. Have a go at the questions below. If you're finding it tough, just look back at the pages in the section and then have another go. You'll be ready to move on when you can answer all of these questions without breaking sweat.

1) Describe the process of chemical weathering.
2) How do waves erode the coast by hydraulic power?
3) What waves are associated with coastal erosion?
4) Describe how a wave-cut platform is formed.
5) Give an example of one type of mass movement.
6) Are headlands made of more or less resistant rock?
7) Describe how erosion can turn a crack in a cliff into a cave.
8) Explain how a stack is formed. Name an example.
9) What is a cove?
10) By what process is material transported along coasts?
11) What is deposition?
12) What waves are associated with coastal deposition?
13) Where is a beach formed on a coast?
14) Why is a sand beach flatter and wider than a shingle beach?
15) Where do spits form? Name an example.
16) Why can't cracks, caves and arches be seen on a map?
17) How are cliffs shown on a map?
18) On maps, what do speckles on top of yellow shading tell you?
19) a) Name a coastal area you have studied which has erosional and depositional landforms.
    b) Name one erosional landform and one depositional landform in that area.
20) Describe one effect of global warming that's causing sea level to rise.
21) Give two economic impacts of coastal flooding and erosion.
22) Give two social impacts of coastal flooding and erosion.
23) a) For a coastal area you have studied, explain why the risk of coastal flooding is becoming greater.
    b) Describe a political impact of coastal flooding in that area.
24) a) Give two main reasons for rapid erosion along a named coastline.
    b) Describe three impacts on people's lives that are caused by erosion of the same coastline.
    c) Describe an environmental impact caused by erosion of the same coastline.
25) Describe the difference between hard engineering and soft engineering coastal management strategies.
26) Explain a disadvantage of using groynes as a coastal defence.
27) a) Name two soft engineering strategies.
    b) Give one benefit of each strategy.
28) a) Give two examples of hard engineering strategies used along a named coastline.
    b) Describe two problems caused by the use of hard engineering strategies along the same coastline.
29) a) Describe how two organisms are adapted to living in a named coastal habitat.
    b) Describe two strategies for dealing with conflicts between land use and conservation in this coastal habitat.

Population Growth

The population change section — putting the 'pop' in 'popular'. Studying population change is amazing — you'd have to be stark raving mad to not agree with me. Only one way to judge for yourself though...

The World's Population is Growing Rapidly

1) The graph shows world population for the years 1500-2000 — it's been increasing and is still increasing today.
2) The population of the world is increasing at an exponential rate — it's growing faster and faster.
3) There are two things that affect the population size of the world:

> Birth rate — the number of live babies born per thousand of the population per year.
>
> Death rate — the number of deaths per thousand of the population per year.

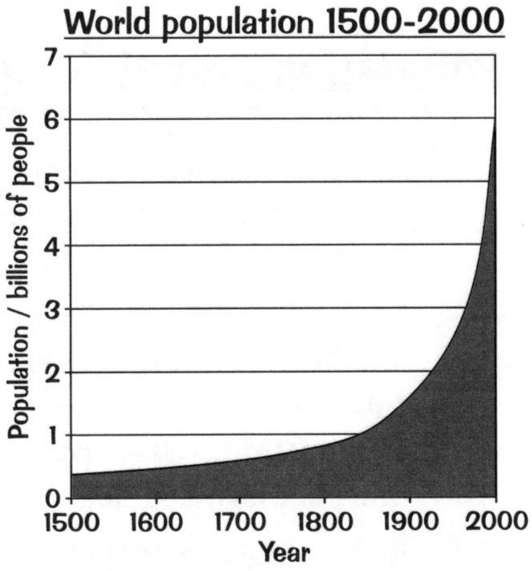

4) When the birth rate is higher than the death rate, more people are being born than are dying, so the population grows — this is called the natural increase.
5) It's called the natural decrease when the death rate's higher than the birth rate.
6) The population size of a country is also affected by migration — the movement of people from one area to another area (see page 96).

Countries go Through Five Stages of Population Growth

1) Birth rates and death rates differ from country to country. This means that population growth is faster in some countries than others.
2) Population growth also changes within a country over time (it can get faster or slower).
3) Countries go through five different stages of population growth.
4) These stages are shown by the Demographic Transition Model (DTM):

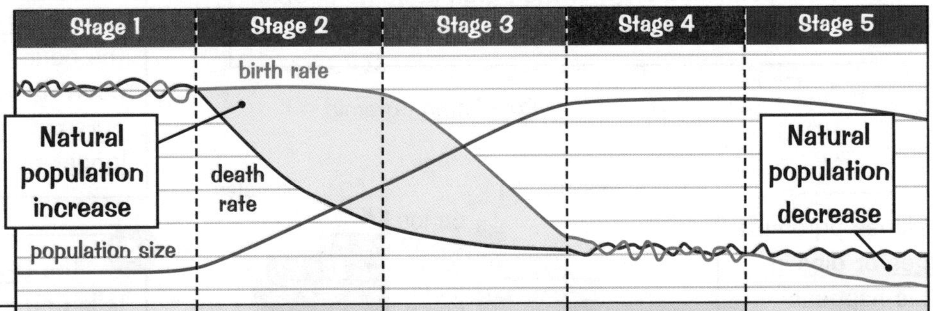

	Stage 1	Stage 2	Stage 3	Stage 4	Stage 5
Birth rate	High and fluctuating	High and steady	Rapidly falling	Low and fluctuating	Slowly falling
Death rate	High and fluctuating	Rapidly falling	Slowly falling	Low and fluctuating	Low and steady
Population growth rate	Zero	Very high	High	Zero	Negative
Population size	Low and steady	Rapidly increasing	Increasing	High and steady	Slowly falling
Example countries	No countries, some tribes in Brazil	Gambia	Egypt	UK, USA	Japan

You know what — this world population is growing on me...

DTM — it stands for Done Too Much (work that is). Sigh, it's a pretty useful thing to know about when you're studying population change though, and it's not too tough to learn, so copy it out and get it imprinted on your memory.

Population Growth and Structure

The population fun's not over yet — you've still got 10 pages to go and the exciting worlds of population pyramids, overpopulation and ageing populations to cover. And then there's migration. I could just pee I'm so excited.

Population Growth is Linked to How Developed a Country is

1) As countries become more developed, birth and death rates change, which affects the population growth. E.g. as a country develops healthcare improves, which leads to a drop in death rate and faster population growth.
2) So as countries become more developed the population changes and the country moves through the stages of the DTM.
3) This means poorer, less developed countries are in the earlier stages of the DTM (2-3) — population growth rate is high because birth rates are high and death rates are beginning to fall.
4) Richer, more developed countries are in the later stages of the DTM (4-5) — they usually have low or negative population growth because birth rates and death rates are low.

Why birth and death rates change as a country becomes more developed is covered on the next page.

A Country's Population Structure Changes as it Develops

The population structure of a country is how many people there are of each age group in the population, and how many there are of each sex. Population structure is shown using population pyramids. Population structure differs from country to country and it changes as countries become more developed. But before you get into the details, you should understand population pyramids. You can learn a lot about a country from its population pyramid. For example:

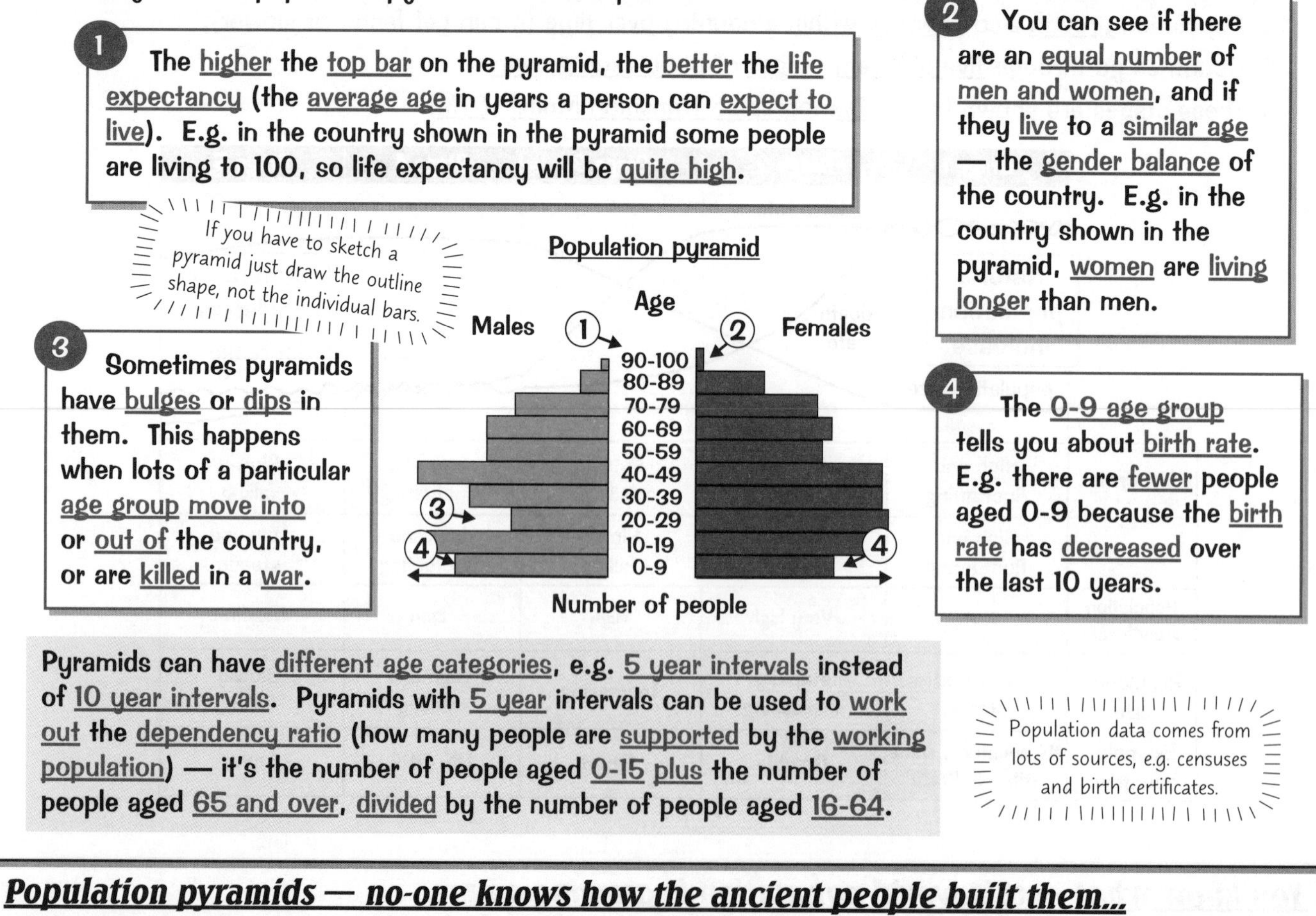

1 The higher the top bar on the pyramid, the better the life expectancy (the average age in years a person can expect to live). E.g. in the country shown in the pyramid some people are living to 100, so life expectancy will be quite high.

2 You can see if there are an equal number of men and women, and if they live to a similar age — the gender balance of the country. E.g. in the country shown in the pyramid, women are living longer than men.

If you have to sketch a pyramid just draw the outline shape, not the individual bars.

3 Sometimes pyramids have bulges or dips in them. This happens when lots of a particular age group move into or out of the country, or are killed in a war.

4 The 0-9 age group tells you about birth rate. E.g. there are fewer people aged 0-9 because the birth rate has decreased over the last 10 years.

Pyramids can have different age categories, e.g. 5 year intervals instead of 10 year intervals. Pyramids with 5 year intervals can be used to work out the dependency ratio (how many people are supported by the working population) — it's the number of people aged 0-15 plus the number of people aged 65 and over, divided by the number of people aged 16-64.

Population data comes from lots of sources, e.g. censuses and birth certificates.

Population pyramids — no-one knows how the ancient people built them...

I've not been to Egypt, but that population pyramid doesn't look like a pyramid to me. It's all dumpy, like it's suffering from pyramid muffin tops. Oh well, make sure you know what they are, whether they're pyramid shaped or not.

Population Growth and Structure

I've never seen so many pyramids in all my life. There's fat and thin, but none made of chocolate and nuts.

There are Many Reasons Why Population Growth and Structure Change

As a country develops and moves through the stages of the DTM its birth and death rates change. This causes the population growth rate, structure and pyramid to change too:

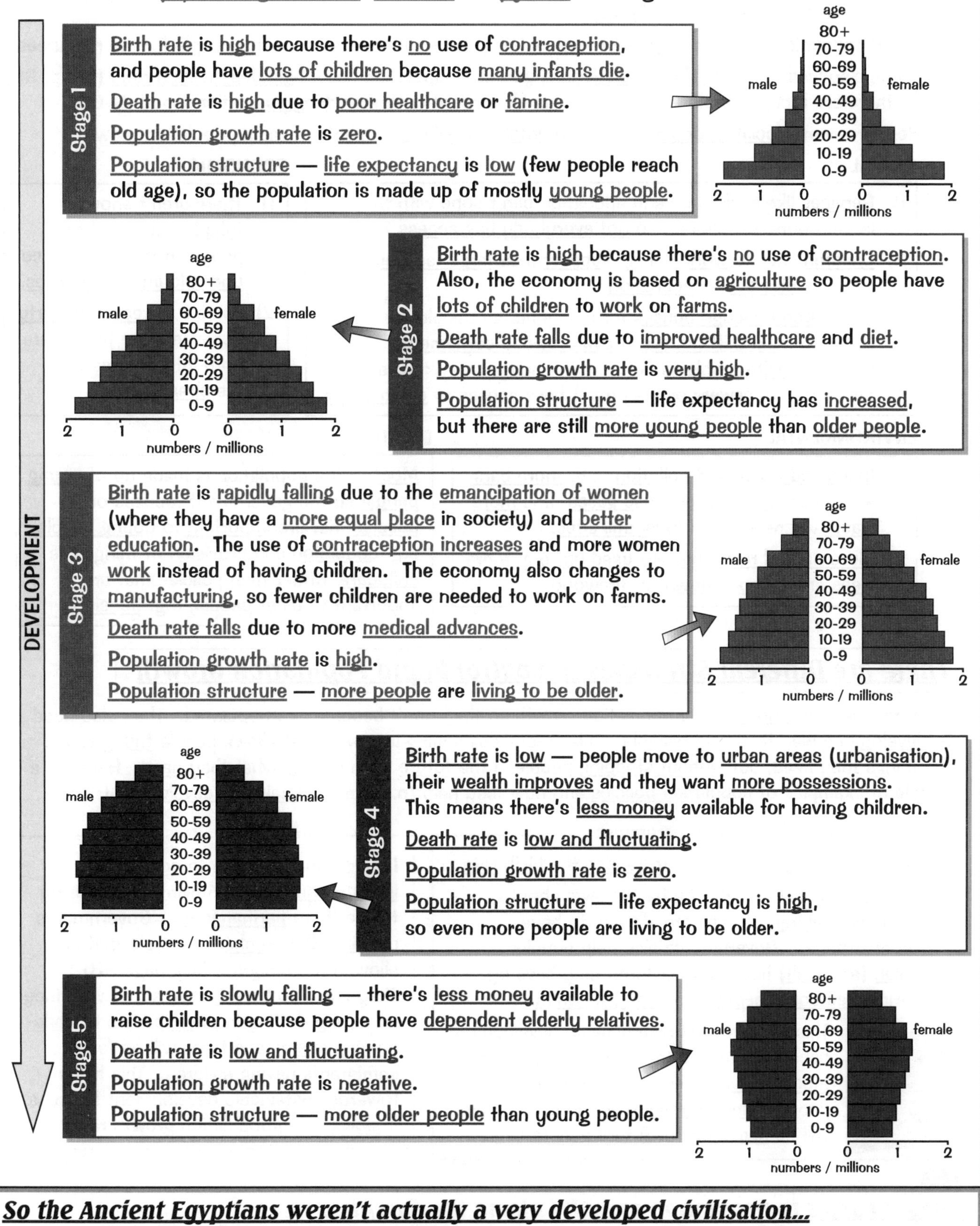

So the Ancient Egyptians weren't actually a very developed civilisation...

There are a fair few words on this page but they're all important. Check you know the reasons why birth rate, death rate, population growth and population structure change as countries become more developed.

Managing Rapid Population Growth

Let's be clear here, rapid population growth doesn't mean everyone is getting taller and fatter. Read on...

Rapid Population Growth has many Negative Impacts

1) Rapid population growth is most likely to happen in poorer countries (such as Gambia) in Stages 2 and 3 of the DTM. They have a high birth rate and a falling death rate, causing a high population growth rate.
2) Rapid population growth can cause overpopulation (when there are too many people for the resources).
3) It can also lead to a youthful population. A youthful population has a high dependency ratio (see p. 90) — there are lots of people under 15 that are dependent on the working population (aged 15–64).
4) Here are some social, economic, environmental and political impacts of rapid population growth:

SOCIAL

1) Services like healthcare and education can't cope with the large, young population, so not everybody has access to them.
2) Children have to work to help support their large families, so they miss out on education.
3) There aren't enough houses for everyone, so people are forced to live in makeshift houses in overcrowded settlements. This leads to health problems because the houses aren't always connected to sewers or they don't have access to clean water.

ECONOMIC

1) There aren't enough jobs for the number of people in the country, so unemployment increases.
2) There's increased poverty because more people are born into families that are already poor.

ENVIRONMENTAL

1) Increased waste and pollution, e.g. more cars will release more greenhouse gases, and more waste will need to go to landfill sites.
2) More natural resources are used up, e.g. more trees are chopped down for firewood.

POLITICAL

Most of the population is made up of young people so the government focuses on policies that are important to young people, e.g. education and provision of things like childcare, rather than policies that are important to older people, e.g. pensions.

There are Different Strategies to Control Rapid Population Growth

Countries need to control rapid population growth so they don't become overpopulated. They also need to develop in a way that's sustainable. This means developing in a way that allows people today to get the things they need, but without stopping people in the future from getting what they need. Here are a couple of examples of population policies and how they help to achieve sustainable development:

Birth control programmes

Birth control programmes aim to reduce the birth rate. Some governments do this by having laws about how many children couples are allowed to have (see next page). Governments also help couples to plan (and limit) how many children they have by offering free contraception and sex education.

This helps towards sustainable development because it means the population won't get much bigger. There won't be many more people using up resources today, so there will be some left for future generations.

Immigration laws

Immigration laws aim to control immigration (people moving to a country to live there permanently). Governments can limit the number of people that are allowed to immigrate (see page 98). They can also be selective about who they let in, e.g. letting in fewer people of child-bearing age means there will be fewer immigrants having children. This helps towards sustainable development because it slows down population growth rate.

When it comes to population growth I blame the parents...

It's not always a case of 'the more the merrier' — rapid population growth can cause problems. That's where strategies to control population growth come in. Check that you know them and how they relate to sustainable development.

Managing Population Growth — Case Studies

This section would be sad and lonely without case studies...

China has a Strict Birth Control Programme

1) China has the largest population of any country in the world — over 1.3 billion.
2) Different policies have been used to control rapid population growth — the most important is the 'one-child policy' introduced in 1979. This means that all couples are very strongly encouraged to have only one child.
3) Couples that only have one child are given benefits like longer maternity leave, better housing and free education for the child. Couples that have more than one child don't get any benefits and are also fined part of their income.
4) Over the years, the policy has changed so there are some exceptions:
    - In some rural areas, couples are allowed to have a second child if the first is a girl, or has a physical disability. This is because more children are still needed to work on farms in rural areas.
    - If one of the parents has a disability or if both parents are only children, then couples are allowed to have a second child. This is so there are enough people to look after the parents.

Effectiveness

1) The policy has prevented up to 400 million births. The fertility rate (the average number of children a woman will have in her life) has dropped from 5.7 in 1970 to around 1.8 today.
2) Some people think that it wasn't just the one-child policy that slowed population growth. They say older policies about leaving longer gaps between children were more effective, and that Chinese people want fewer children anyway as they've become more wealthy.

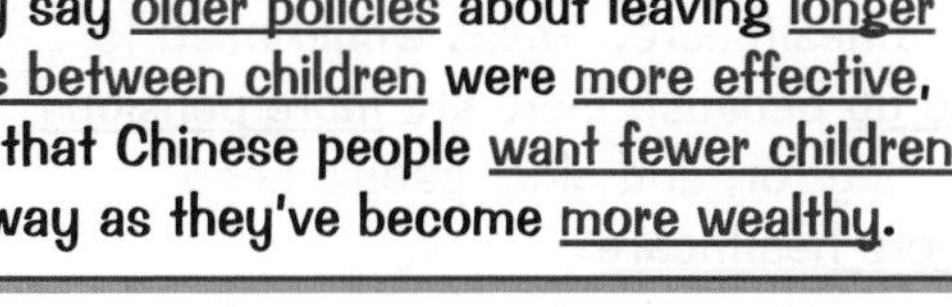

China's one-child policy helps towards sustainable development — the population hasn't grown as fast (and got as big) as it would have done without the policy, so fewer resources have been used.

Indonesia has Tried to Tackle the Problems of Rapid Population Growth

1) Indonesia is a country made up of thousands of islands. It has the fourth largest population of any country in the world — over 240 million.
2) The population isn't distributed evenly — most people (around 130 million) live on the island of Java.
3) This has led to social and economic problems (see the previous page) on the densely populated islands, e.g. a lack of adequate services and housing as well as unemployment and poverty.
4) The Indonesian Government started a policy in the 1960s called the transmigration policy, which aims to reduce the impacts of population growth.
5) Millions of people have been moved from the densely populated islands like Java, to the less densely populated islands like Sumatra.

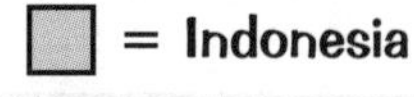

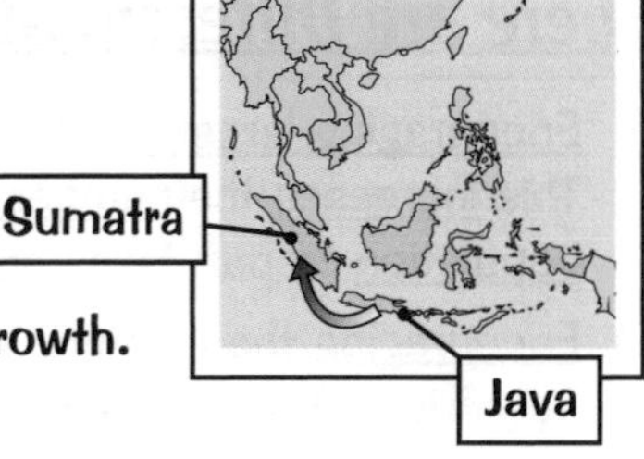

Effectiveness

1) Millions of people have been moved, but the population still isn't much more evenly distributed.
2) Not all the people who were moved escaped poverty — either they didn't have the skills to farm the land, or the land was too poor to be farmed on their new island.
3) Lots of people were moved to land that was already occupied by native people. This created a new problem — conflict between the natives and the migrants.

Indonesia's transmigration policy hasn't helped towards sustainable development because it only reduces the impacts of population growth — the population is still getting much bigger.

Someone should tell the pandas they can have as many babies as they like...

In some cases, governments have had to take pretty drastic action to deal with rapid population growth.
Just try and imagine what you'd say if your local MP knocked on your door and told you to move to another island...

Ageing Populations

An ageing population is one that has a high proportion of older people. Ageing populations can face some sticky economic and social problems. They also tend to consume a high quantity of teacakes.

An Ageing Population Impacts on Future Development

The population structure of an ageing population has more older people than younger people because few people are being born, and more people are surviving to old age.

age
80+
70-79
60-69
50-59
40-49
30-39
20-29
10-19
0-9
male
female
2 1 0 0 1 2
numbers / millions

Countries with an ageing population are usually the richer countries in Stage 5 of the DTM (see pages 89 and 91).

Older people (over 65) are supported by the working population (aged 16-64) — they're dependent on them. So in a country with an ageing population there's a higher proportion of people who are dependent. This has economic and social impacts, which can affect a country's future development:

ECONOMIC

1) The working population pay taxes, some of which the government use to pay the state pensions of older people, and to pay for services like retirement homes and healthcare. Taxes would need to go up because there are more pensions to pay for, and older people need more healthcare.
2) The economy of the country would grow more slowly — less money is being spent on things that help the economy to grow, e.g. education and business, and more money is being spent on things that don't help the economy to grow, e.g. retirement homes.

SOCIAL

1) Healthcare services are stretched more because older people need more medical care.
2) People will need to spend more time working as unpaid carers for older family members. This means that the working population have less leisure time and are more stressed and worried.
3) People may have fewer children because they can't afford lots of children when they have dependent older relatives. This leads to a drop in birth rate.
4) The more old people there are, the lower the pension provided by the government will be. People will have to retire later because they can't afford to get by on a state pension.

There are Different Strategies to Cope with an Ageing Population

1) Encouraging larger families, e.g. in Italy women are offered cash rewards to have more children. This increases the number of young people — when they start work there will be a larger working population to pay taxes and support the ageing population.
2) Encouraging the immigration of young people from other countries. This increases the working population so there are more people paying taxes to support the ageing population.

These strategies don't help towards sustainable development because they increase the population size.

3) Raising the retirement age — people stay in work longer and contribute to state pensions and personal pensions for longer. They will also claim the state pension for less time.
4) Raising taxes for the working population — this would increase the amount of money available to support the ageing population.

These strategies help towards sustainable development because they help to reduce the impacts of an ageing population, without increasing the population size.

Being old is not a crime but my great-aunt has knitted some criminal jumpers...

'Live long enough to be a burden on your children' — I thought it was just a phrase... Learn about the social and economic impacts an ageing population can have, and the strategies governments have come up with to deal with them.

Ageing Populations — Case Study

Like most wealthy and developed countries around the world, the UK has an ageing population. This case study's got lots of juicy statistics for you to read about, so get swotting.

The UK's Population is Ageing

In 2005, 16% of the population of the UK was over 65. By 2041 this could be 25%.

The Ageing Population is Caused by Increasing Life Expectancy and Dropping Birth Rate

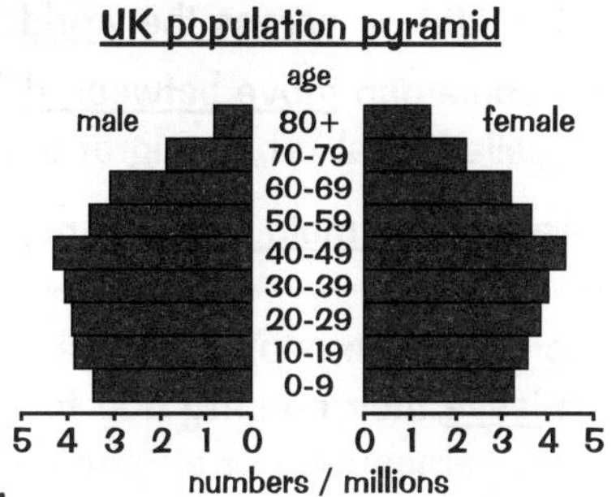

1) People are living longer because of advances in medicine and improved living standards. Between 1980 and 2006 life expectancy rose 2.6 years for women and 6.4 years for men — it's currently 81.5 for women and 77.2 for men. This means the proportion of older people in the population is going up.
2) Lots of babies were born in the 1940s and 1960s — periods called 'baby booms'. Those born in the 1940s are retiring now, creating a 'pensioner boom'.
3) Since the 1970s, the number of babies born has fallen. With fewer young people in the population the proportion of older people goes up.

The UK's Ageing Population Causes a few Problems

1) More elderly people are living in poverty — the working population isn't large enough to pay for a decent pension, and many people don't have other savings.
2) Even though the state pension is low the government is struggling to pay it. The taxes paid by people in work aren't enough to cover the cost of pensions and as the population ages the situation is getting worse.
3) The health service is under pressure because older people need more medical care than younger people. For example, in 2005 the average stay in hospital for people over 75 was 13 nights, but for the whole of the UK the average stay was only 8 nights.

The UK Government has Strategies to Cope with the Ageing Population

1) Raise the retirement age — the retirement age in the UK is currently 65 for men and 60 for women. This is going to change in stages, so that by 2046 it will be 68 for everyone. People will have to work for longer, so there will be more people paying tax and fewer claiming a pension.
2) Encourage immigration of young people to the UK — the UK has allowed immigration of people from countries that joined the EU in 2004. Around 80% of immigrants from new EU countries in 2004 were 34 or under. This increases the number of people paying taxes, which helps to pay for the state pension and services.
3) Encourage women to have children — working family tax credits support women (and men) who go back to work after their children are born. This makes it more affordable for couples to have children.
4) Encourage people to take out private pensions — the government gives tax breaks for some types of private pension. With private pensions, people won't be so dependent on the state pension.

We Don't Know if the Strategies have Worked Yet

It's too early to tell if government strategies are working. Even if they do have some effect it's likely that future generations will have to work longer and rely on their families to support them in old age.

Ageing population impact — sales of toilet roll covers are through the roof...

This case study is perfect exam fodder. Nothing makes an examiner's eyes spin like fruit machines than real-life examples. Memorise the facts and figures from this page and you'll be all set up to get top marks in your exam.

Migration

Migration is the movement of people from one area to another area. However, they don't just do it for the fun of it — this page is all about the reasons why, as well as some good ol' impacts.

People Migrate Within Countries and To Different Countries

1) When people move into an area, it's called immigration. The people are called immigrants.
2) When people exit an area, it's called emigration. The people are called emigrants.
3) People can move to different countries — this is known as international migration. It might be across the world, or just a few miles over a border.
4) People can move between different regions within countries, e.g. from the countryside to a city (called rural-urban migration). This is known as internal migration.

There are two main types of migrant:

Refugees are people who've been forced to leave their country due to things like war, persecution or a natural disaster, e.g. thousands of refugees migrated to escape the war in Kosovo in 1999.

Economic migrants are people who move voluntarily from poorer places to richer places looking for jobs or higher wages, e.g. from Mexico to the USA. They often migrate so they can earn more money and then send some back to family in their country of origin.

Migration Happens Because of Push and Pull Factors

The reasons a person migrates can be classified as either push or pull factors:

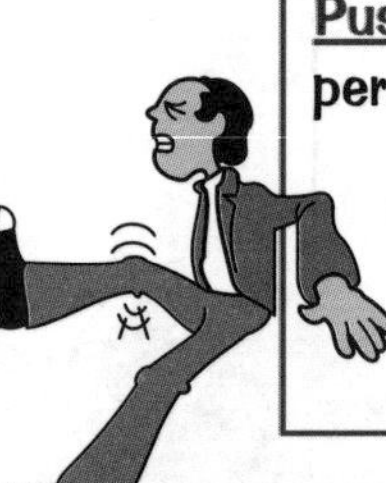

Push factors are negative things about a person's place of origin (where they originally lived) that make them want to leave. They're usually things like not being able to find a job, poor living conditions, war or a natural disaster in their country of origin.

Pull factors are positive things about a person's destination that attract them to the destination. They're usually things such as job opportunities or a better standard of living.

Migration Has Positive and Negative Impacts

Migration has impacts on both the source country (where they come from) and the receiving country (where they're going to):

	Positive impacts	Negative impacts
Source country	Reduced demand on services, e.g. schools and hospitals. Money is sent back to the source country by emigrants.	Labour shortage — it's mostly people of working age that emigrate. Skills shortage — sometimes it's the more highly educated people that emigrate. Ageing population — there's a high proportion of older people left.
Receiving country	Increased labour force — young people immigrate to find work. Migrant workers pay taxes that help to fund services.	Locals and immigrants compete for jobs — this can lead to tension and even conflict. Increased demand for services, e.g. overcrowding in schools and hospitals. Not all the money earnt by immigrants is spent in the destination country — some is sent back to their country of origin.

Umm-igration is when you don't know whether you're coming or going...

Migration sounds like a rough business, people being pushed and pulled all over the place. Remember that migration affects both the place the people leave, and the place they go to, and that the effects can be positive or negative.

International Migration

It seems that an awful lot of people are on the move, and we're not just talking a quick trip into town. Here are a couple of examples to spice up your life...

Case Study 1 — There are Economic Migrations From Poland To the UK

People who come from a country in the EU can live and work in any other EU country. In 2004, ten eastern European countries joined the EU. Since then, people from these countries have been moving to other EU countries. More than half a million people from Poland came to the UK between 2004 and 2007.

There were push and pull factors for why people left Poland and came to the UK:

Push factors from Poland (in 2004):

1) High unemployment — around 19%.
2) Low average wages — about one third of the average EU wage.
3) Housing shortages — just over 300 dwellings for every 1000 people.

Pull factors to the UK:

1) Ease of migration — the UK allowed unlimited migration in 2004 (it was restricted in some other EU countries).
2) More work and higher wages — wages in the UK were higher and there was a big demand for tradesmen, e.g. plumbers.
3) Good exchange rate — the pound was worth a lot of Polish currency, so sending a few pounds back to Poland made a big difference to family at home.

IMPACTS IN POLAND

1) Poland's population fell (by 0.3% between 2003 and 2007), and the birth rate fell as most people who left were young.
2) There was a shortage of workers in Poland, slowing the growth of the economy.
3) The Polish economy was boosted by the money sent home from emigrants — around €3 billion was sent to Poland from abroad in 2006.

IMPACTS IN THE UK

1) The UK population went up slightly.
2) Immigration boosted the UK economy, but a lot of the money earned in the UK was sent home.
3) New shops selling Polish products opened to serve new Polish communities.
4) Many Poles are Catholic so attendance at Catholic churches went up.

Case Study 2 — Refugees Migrate To the EU

Huge numbers of people migrate from Africa to the EU. For example, by crossing the Mediterranean sea to Spain — in 2001, 45 000 emigrants from Africa were caught and refused entry to Spain.

Many of these migrants are refugees (see the previous page) from wars in central and western African countries. For example, more than 2 million people were forced from their homes because of the civil war in Sierra Leone (in West Africa) between 1991 and 2002.

There are only push factors for African refugees of war — people flee the countries because of the threat of violence or death during the wars.

Here are some of the impacts:

IMPACTS IN AFRICAN COUNTRIES

1) The working population is reduced so there are fewer people contributing to the economy.
2) Families become separated when fleeing from wars.

IMPACTS IN SPAIN

1) Social tension between immigrants and Spaniards.
2) More unskilled workers in Spain, which has filled gaps in the labour market.
3) Average wages for unskilled jobs have fallen because there are so many people who want the jobs.
4) The birth rate has increased because there are so many young immigrants.

Economic migration — the money's getting away...

Funnily enough, lots of people really want to come and live in the UK. Sure Australia's got better weather and the trains in Italy always run on time, but we've got the Queen, weak tea and the jewel in the crown... Blackpool.

Managing International Migration

International migration causes population change in a country, so some countries try to manage it. Make sure you know some examples of how it can be managed — and a case study too.

There are Different Ways to Manage International Migration

Different countries manage international migration in different ways. Here are a few examples:

POINTS-BASED SYSTEMS

1) Points-based systems let countries choose who they want to let in.
2) People who want to move are given points for things like age, education, work experience and whether they speak the language.
3) Only those with enough points are allowed in, so in theory only the most skilled immigrants who'll adapt well are allowed to enter.

Australia, New Zealand and Canada use systems like this.

6.1

8.5

BORDER CONTROL

LIMITS AND TARGETS

1) Limits and targets are set by some countries to make sure they don't let in too many or too few immigrants (although having too few is less likely to be a problem for most countries).
2) The limits are set by looking at things like how many jobs are available and public opinion.
3) If the limit has already been reached that year, no one else is allowed in.

CONTROLLING ILLEGAL IMMIGRATION

1) Lots of countries, especially richer countries, have problems with people entering illegally or staying after they should have left (e.g. at the end of a holiday).
2) In many countries illegal immigrants can be arrested and forced to leave the country, e.g. Italy fines and deports illegal immigrants, and people can go to jail for knowingly housing an illegal immigrant.

Case Study — The UK has now Changed How it Manages Immigration

Economic migration from Poland to the UK has many impacts (see Case Study 1 on page 97). As a result, the UK has had to change how it manages immigration.

1) Immigrants from Poland entering the UK aren't limited in number, but they do have to register under the Worker Registration Scheme if they want to work in the UK.
2) This lets the UK Border Agency monitor how many people are coming into the country, what type of work they're doing and the effect this is having on the UK economy.
3) The large number of Polish immigrants entering the UK led to some complaints — some people thought the resources in the UK wouldn't be able to cope with all the new people, e.g. there wouldn't be enough jobs or housing to go around.
4) In response to this, the Government tightened the control of migration from some of the newer EU states. For example, immigrants from the two newest EU states, Bulgaria and Romania, have to get permission from the Home Office to work in the UK (and this is only granted for certain types of jobs).

At least no one has stopped yummy Polish sausage being sold in Tesco....

Managing immigration is a tricky issue. There are lots of ways to do it and none of them pleases everybody — think how long politicians spend arguing about it. Don't argue about learning this page though — just get it in your noggin.

Revision Summary for Section 8

That's another smashing section under your belt — congratulations, oh studious one. And here's a delightful array of questions so you can check you've taken it all in. If you'd care to begin...

1) What are the two things that affect the population size of the world?
2) Under what circumstances does natural increase happen to a population?
3) What happens to death rate at Stage 2 of the DTM?
4) What happens to birth rate at Stage 3 of the DTM?
5) Are richer countries or poorer countries more likely to be in the early stages of the DTM?
6) What is the population structure of a country?
7) Give one reason why birth rate is high during Stage 1 of the DTM.
8) Describe how changes in the economy affect the population growth rate.
9) Give one reason why the birth rate rapidly falls during Stage 3 of the DTM.
10) Briefly describe the population structure of a country in Stage 5 of the DTM.
11) Give two social impacts and one economic impact of rapid population growth.
12) Describe what it means for a country to develop in a way that's sustainable.
13) Give an example of a strategy a country could use to control rapid population growth.
14) a) Name a population control policy used in a country you have studied.
    b) Describe the policy.
    c) Explain whether you think the policy has been effective.
15) What is an ageing population?
16) Give two causes of an ageing population.
17) Give one economic impact of an ageing population.
18) Give two social impacts of an ageing population.
19) Describe one strategy to cope with an ageing population.
20) a) Name a country that has an ageing population and describe what caused it.
    b) For the country you've named, describe one problem caused by the ageing population.
    c) Give one strategy used by the country to influence population change.
    Explain how it helps, and comment on its sustainability.
21) Define 'migration'.
22) What's it called when a person moves into an area?
23) a) Explain what 'pull factors' are.
    b) Give an example of a pull factor.
24) Give one negative impact of migration on a source country.
25) a) Describe an example of international migration.
    b) Describe the push and pull factors.
    c) Give one impact on the source country, and one impact on the receiving country.
26) Describe one way in which international migration can be managed.
27) For a country you have studied, describe a strategy used to try to control international migration.

Urbanisation

Urban areas (towns and cities) are the place to be... or so I'm told by my mate who tucks his tracksuit bottoms into his socks. And he's right — people are upping sticks and moving to cities by the bucketload.

Urbanisation is Happening Fastest in Poorer Countries

Urbanisation is the growth in the proportion of a country's population living in urban areas.
It's happening in countries all over the world — more than 50% of the world's population currently live in urban areas (3.4 billion people) and this is increasing every day.
But urbanisation differs between richer and poorer countries:

1) Most of the population in richer countries already live in urban areas, e.g. more than 80% of the UK's population live in urban areas.
2) Not many of the population in poorer countries currently live in urban areas, e.g. around 25% of the population of Bangladesh live in urban areas.
3) Most urbanisation that's happening in the world today is going on in poorer countries and it's happening at a fast pace.

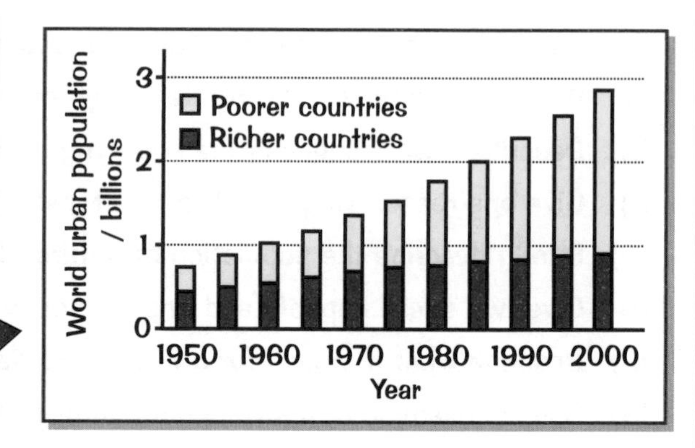

Urbanisation is Caused By Rural-urban Migration...

Rural-urban migration is the movement of people from the countryside to the cities.
Rural-urban migration causes urbanisation in richer and poorer countries.
The reasons why people move are different in poorer and richer countries though.

Here are a few reasons why people in poorer countries move from rural areas to cities:

1) There's often a shortage of services (e.g. education, access to water and power) in rural areas. Also, people from rural areas sometimes believe that the standard of living is better in cities (even though this often turns out not to be the case).
2) There are more jobs in urban areas. Industry is attracted to cities because there's a larger workforce and better infrastructure than in rural areas.
3) In rural areas some people are subsistence farmers. This means they grow food to feed their family and sell any extra to make a small income. Poor harvests and crop failures can mean they make no income and even risk starvation.

Here are a couple of reasons why people in richer countries move from rural areas to cities:

1) Most urbanisation in rich countries occurred during the Industrial and the Agricultural Revolutions (18th and 19th centuries) — machinery began to replace farm labour in rural areas, and jobs were created in new factories in urban areas. People moved from farms to towns for work.
2) In the late 20th century, people left run-down inner city areas and moved to the country. But people are now being encouraged back by the redevelopment of these areas.

... And Good Healthcare and a High Birth Rate in Cities

It's normally young people that move to cities to find work. These people have children in the cities, which increases the proportion of the population living in urban areas. Also, better healthcare in urban areas means people live longer, again increasing the proportion of people in urban areas.

Dick Whittington and His Cat — the classic rural-urban migration case study...

Nothing too difficult on this page — richer countries have a high percentage of their population in urban areas, but urbanisation in poorer countries is happening faster than you can say 'oh my giddy aunt, that's some rapid urbanisation'.

Urban Land Use

Every city is different, but they all have dodgy run-down parts and posh housing areas. Obviously you should use more formal terms to describe them in the exam — amazingly enough, they're all listed below...

A City can be Split into Four Main Parts

Cities are usually made up of four parts — each part has a different land use (e.g. housing or industrial). The land use of each part stays fairly similar from city to city, but it can differ a bit (see below). The diagram below is a view from above of a typical city — it shows roughly where the four parts are:

CBD

This is the central business district. It's usually found right in the centre of a city. It's the commercial centre of the city with shops and offices.

The inner city

This part is found around the CBD. It has a mix of poorer quality housing (like high-rise tower blocks) and older industrial buildings.

This is just a model — no city looks exactly like this.

The suburbs

These are housing areas found towards the edge of the city.

The rural-urban fringe

This is the part right at the edge of a city, where there are both urban land uses (e.g. factories) and rural land uses (e.g. farming).

Land Use is decided by Social, Economic and Cultural Factors

Part of city	Land use	Social factors	Economic factors	Cultural factors
CBD	Businesses, e.g. shops and offices	It's busy and very accessible.	Land is expensive (only businesses can afford it).	It's a centre point for entertainment, e.g. cinemas.
Inner city	Low-class housing and industry	Traditionally small houses were built here near to factories to house workers.	Poorer people who can't afford to commute and can only afford small houses live here.	Ethnic groups live here so they're near to important services, e.g. places of worship.
Suburbs	Medium-class housing	It's less crowded and more pleasant, with less traffic and pollution.	Richer people who can afford to commute and to have big houses live here.	People with families live here due to the space for leisure activities, e.g. BBQs.
Rural-urban fringe	Business parks and high-class housing	It's still accessible for commuters and there's lots of space.	The land is often cheaper here so bigger houses can be built for richer people.	Richer people who like a rural lifestyle and being in reach of the city live here.

The Land Use of the Parts can Differ from City to City

1) Sometimes the land use of each part doesn't match the model above — real cities are all slightly different. For example, in cities in poorer countries, e.g. Rio de Janeiro, there's usually low-cost housing and squatter settlements on the outskirts of cities, but high-class housing in the CBD.
2) The land use of each part of a city can also change over time, for example:
   - In recent years a lot of shopping centres have been built in out-of-town locations in the UK, e.g. Meadowhall Shopping Centre was built on the outskirts of Sheffield in 1990.
   - Inner city tower blocks have been removed and replaced with housing estates on the rural-urban fringe, e.g. this has happened in Birmingham.

Just popping into the rural-urban fringe to do a bit of shopping...

Check that you know the four main parts of a city and the land use in each bit, but remember that the land use isn't the same everywhere — I'm sure city planners do this on purpose, just to make your revision awkward. Some people...

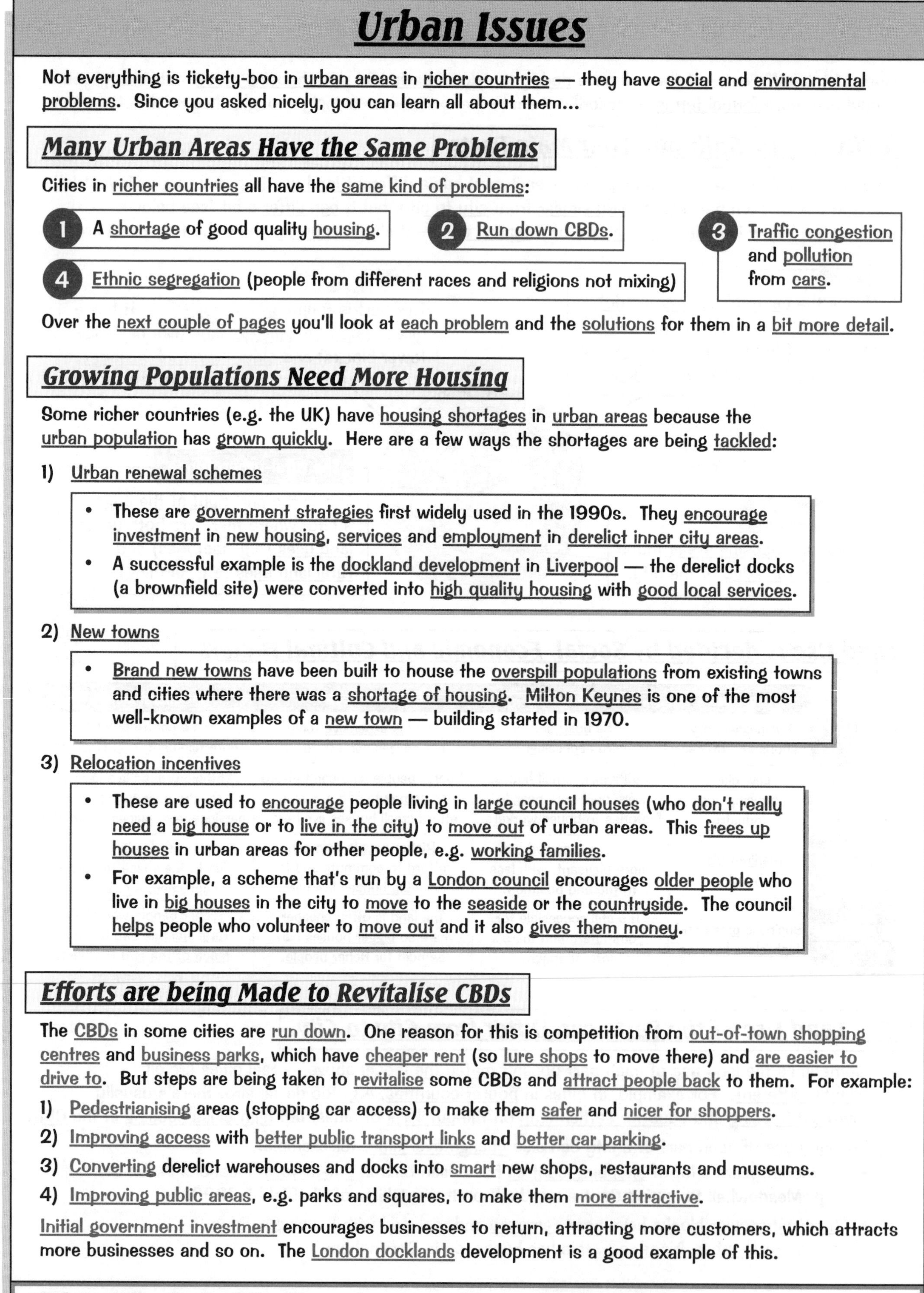

Urban Issues

Not everything is tickety-boo in urban areas in richer countries — they have social and environmental problems. Since you asked nicely, you can learn all about them...

Many Urban Areas Have the Same Problems

Cities in richer countries all have the same kind of problems:

1. A shortage of good quality housing.
2. Run down CBDs.
3. Traffic congestion and pollution from cars.
4. Ethnic segregation (people from different races and religions not mixing)

Over the next couple of pages you'll look at each problem and the solutions for them in a bit more detail.

Growing Populations Need More Housing

Some richer countries (e.g. the UK) have housing shortages in urban areas because the urban population has grown quickly. Here are a few ways the shortages are being tackled:

1) Urban renewal schemes

- These are government strategies first widely used in the 1990s. They encourage investment in new housing, services and employment in derelict inner city areas.
- A successful example is the dockland development in Liverpool — the derelict docks (a brownfield site) were converted into high quality housing with good local services.

2) New towns

- Brand new towns have been built to house the overspill populations from existing towns and cities where there was a shortage of housing. Milton Keynes is one of the most well-known examples of a new town — building started in 1970.

3) Relocation incentives

- These are used to encourage people living in large council houses (who don't really need a big house or to live in the city) to move out of urban areas. This frees up houses in urban areas for other people, e.g. working families.
- For example, a scheme that's run by a London council encourages older people who live in big houses in the city to move to the seaside or the countryside. The council helps people who volunteer to move out and it also gives them money.

Efforts are being Made to Revitalise CBDs

The CBDs in some cities are run down. One reason for this is competition from out-of-town shopping centres and business parks, which have cheaper rent (so lure shops to move there) and are easier to drive to. But steps are being taken to revitalise some CBDs and attract people back to them. For example:

1) Pedestrianising areas (stopping car access) to make them safer and nicer for shoppers.
2) Improving access with better public transport links and better car parking.
3) Converting derelict warehouses and docks into smart new shops, restaurants and museums.
4) Improving public areas, e.g. parks and squares, to make them more attractive.

Initial government investment encourages businesses to return, attracting more customers, which attracts more businesses and so on. The London docklands development is a good example of this.

Sittin' on the dock of the bay — in a very chic new Italian place...

A lack of housing and run down CBDs — it's easy to see why people love cities. To be fair, those city slickers have some good ideas on how to deal with housing shortages and revitalise CBDs — check how many of them you remember.

Urban Issues

If you're hungry for more urban problems and solutions then you've come to the right place. Bon appétit...

Increased Car Use has an Impact on Urban Environments

There are more and more cars on the roads of cities in richer countries. This causes a variety of problems, which can discourage people from visiting and shopping in the city:

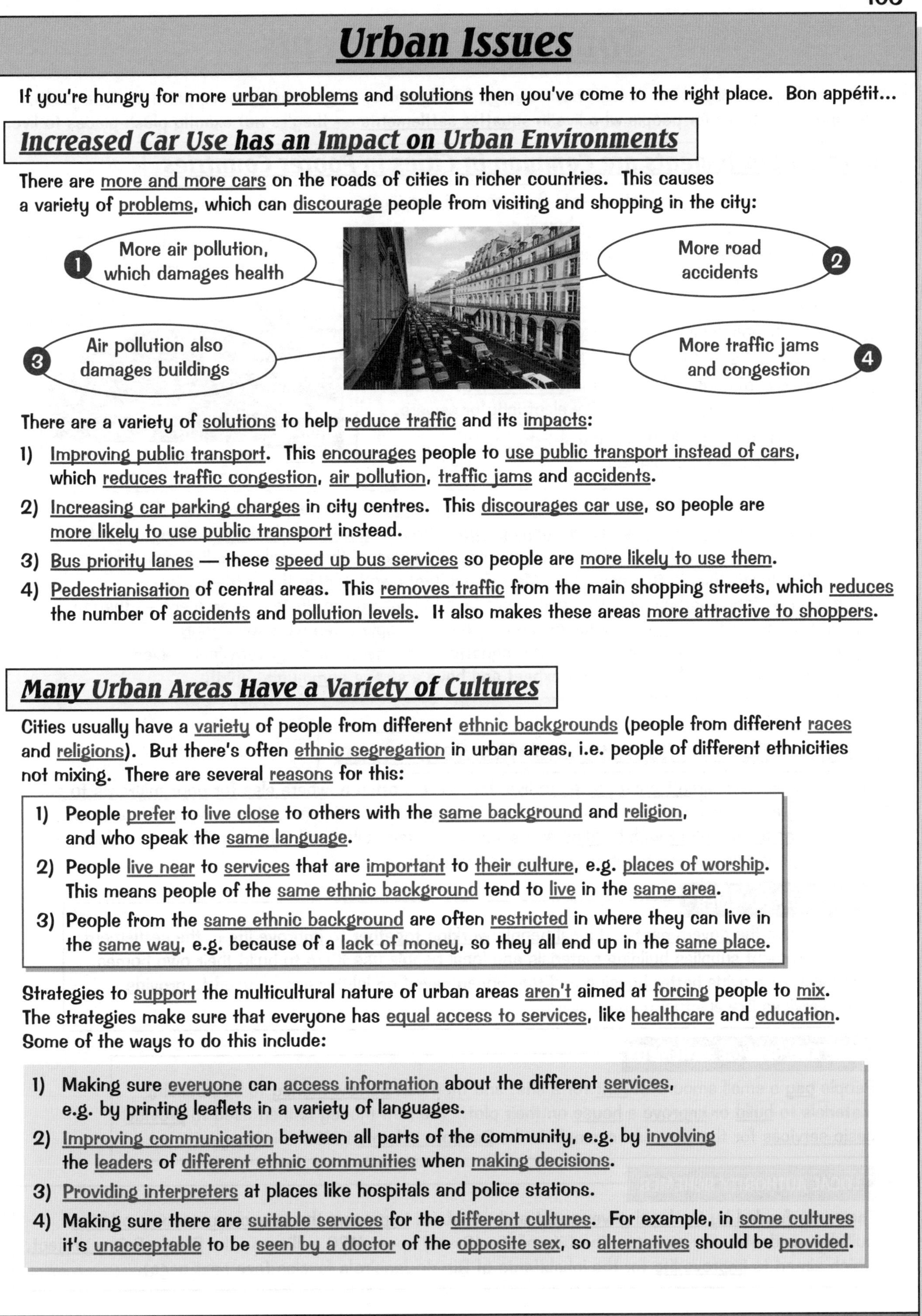

There are a variety of solutions to help reduce traffic and its impacts:

1) Improving public transport. This encourages people to use public transport instead of cars, which reduces traffic congestion, air pollution, traffic jams and accidents.
2) Increasing car parking charges in city centres. This discourages car use, so people are more likely to use public transport instead.
3) Bus priority lanes — these speed up bus services so people are more likely to use them.
4) Pedestrianisation of central areas. This removes traffic from the main shopping streets, which reduces the number of accidents and pollution levels. It also makes these areas more attractive to shoppers.

Many Urban Areas Have a Variety of Cultures

Cities usually have a variety of people from different ethnic backgrounds (people from different races and religions). But there's often ethnic segregation in urban areas, i.e. people of different ethnicities not mixing. There are several reasons for this:

1) People prefer to live close to others with the same background and religion, and who speak the same language.
2) People live near to services that are important to their culture, e.g. places of worship. This means people of the same ethnic background tend to live in the same area.
3) People from the same ethnic background are often restricted in where they can live in the same way, e.g. because of a lack of money, so they all end up in the same place.

Strategies to support the multicultural nature of urban areas aren't aimed at forcing people to mix. The strategies make sure that everyone has equal access to services, like healthcare and education. Some of the ways to do this include:

1) Making sure everyone can access information about the different services, e.g. by printing leaflets in a variety of languages.
2) Improving communication between all parts of the community, e.g. by involving the leaders of different ethnic communities when making decisions.
3) Providing interpreters at places like hospitals and police stations.
4) Making sure there are suitable services for the different cultures. For example, in some cultures it's unacceptable to be seen by a doctor of the opposite sex, so alternatives should be provided.

Perhaps it would also help if the slang term for bus wasn't 'loser cruiser'...

All this stuff seems pretty straightforward but keep going over it until it's all lodged in your noggin. Cover the page and try to remember the problems and the solutions. Don't move on until you can remember them all.

Squatter Settlements

If you think your living room doesn't have enough comfy sofas, or your bathroom's just not big enough, then spare a thought for people who live in squatter settlements — they're not exactly plush places to live.

Squatter Settlements are Common in Cities in Poorer Countries

1) Squatter settlements are settlements that are built illegally in and around the city, by people who can't afford proper housing.
2) Squatter settlements are a problem in many growing cities in poorer countries, e.g. São Paulo (Brazil) and Mumbai (India).
3) Most of the inhabitants have moved to the city from the countryside — they're rural–urban migrants.
4) The settlements are badly built and overcrowded. They often don't have basic services like electricity or sewers.
5) They're called favelas in Brazil and shanty towns or slums in some other places.

Life in a squatter settlement can be hard and dangerous — the people living there don't have access to basic services like clean running water, proper sewers or electricity. They may also lack policing, medical services and fire fighting. Because of these problems, life expectancy is often lower than in the main city. Many inhabitants work within the settlements, e.g. in factories and shops. The jobs aren't taxed or monitored by the government — they're referred to as the informal sector of the economy. People often work long hours for little pay in the informal sector. But squatter settlements often govern themselves more successfully than you might expect and have a strong community spirit.

There are Ways to Improve Squatter Settlements

Squatter settlements aren't great places to live, but there's often nowhere else for poor migrants to go. People living in squatter settlements usually try to improve the settlements themselves. For example, neighbours help each other with building and some have even built small schools. But the residents have little money and can achieve much more with a bit of help:

SELF-HELP SCHEMES

These involve the government and local people working together to improve life in the settlement. The government supplies building materials and local people use them to build their own homes. This helps to provide better housing and the money saved on labour can be used to provide basic services like electricity and sewers.

SITE AND SERVICE SCHEMES

People pay a small amount of rent for a site, and they can borrow money to buy building materials to build or improve a house on their plot. The rent money is then used to provide basic services for the area. An example is the Dandora scheme in Nairobi, Kenya.

LOCAL AUTHORITY SCHEMES

These are funded by the local government and are about improving the temporary accommodation built by residents. For example, the City of Rio (Brazil) spent $120 million on the Favela-Bairro project, which aimed to improve life for the inhabitants of Rio de Janeiro's favelas (see next page).

Some residents even try to win 'Who Wants to be a Millionaire'...

People who live in squatter settlements try to improve the conditions themselves. It's not much like your dad's failed DIY attempts though — forget about putting up wonky shelves, these people are putting a roof over their heads.

Squatter Settlements — Case Study

And now, introducing tonight's case study, all the way from Brazil, lets hear a big hand for the Favela-Bairro Project in Rio de Janeiro. Oh come on, surely you can muster a smile...

The Favela-Bairro Project Helps People in Rio de Janeiro's Favelas

1) Rio de Janeiro is in south east Brazil. It has 600 squatter settlements (favelas), housing one-fifth of the city's population (more than one million people).
2) The Favela-Bairro project started in 1995 and is so successful it's been suggested as a model for redeveloping other squatter settlements.
3) The project involves 253 000 people in 73 favelas, and is being extended to help even more people.
4) 40% of the $300 million funding for the project came from the local authority. The rest was provided by an international organisation called the Inter-American Development Bank.

The Project Includes Social, Economic and Environmental Improvements

1) Social improvements:
   - Daycare centres and after school schemes to look after children while their parents work.
   - Adult education classes to improve adult literacy.
   - Services to help people affected by drug addiction, alcohol addiction and domestic violence.
2) Economic improvements:
   - Residents can now apply to legally own their properties.
   - Training schemes to help people learn new skills so they can find better jobs and earn more.
3) Environmental improvements:
   - Replacement of wooden buildings with brick buildings and the removal of homes on dangerously steep slopes.
   - Widening and paving of streets to allow easier access (especially for emergency services).
   - Provision of basic services such as clean water, electricity and weekly rubbish collection.

Community involvement is one of the most important parts of the project:

- Residents choose which improvements they want in their own favela, so they feel involved.
- Neighbourhood associations are formed to communicate with the residents and make decisions.
- The new services are staffed by residents, providing income and helping them to learn new skills.

The Favela-Bairro Project has been Very Successful

1) The standard of living and health of residents have improved.
2) The property values in favelas that are part of the programme have increased by 80–120%.
3) The number of local businesses within the favelas has almost doubled.

Her name is Rio and she dances on the recently widened and paved roads...

If you get a squatter settlement case study question in the exam, you need to dazzle the examiner with lots of facts and figures. It's no good just saying that favelas aren't very swish. It might be true but it won't get you any marks.

Urbanisation — Environmental Issues

It may be faster than a speeding bullet, but rapid urbanisation in poorer countries brings a whole host of environmental problems... which is a bit of a shame because that means there are lots for you to learn.

Rapid Urbanisation and Industrialisation Affect the Environment

Rapid urbanisation and industrialisation (where the economy of a country changes from being based on agriculture to manufacturing) can cause a number of environmental problems:

1) Waste disposal problems — people in cities create a lot of waste. This can damage people's health and the environment, especially if it's toxic and not disposed of properly.
2) More air pollution — this comes from burning fuel, vehicle exhaust fumes and factories.
3) More water pollution — water carries pollutants from cities into rivers and streams. For example, sewage and toxic chemicals from industry can get into rivers which causes serious health problems. Wildlife can also be harmed.

Waste Disposal is a Serious Problem in Poorer Countries

In richer countries, waste is disposed of by burying it in landfill sites, or by burning it. The amount of waste is also reduced by recycling schemes. Poorer countries struggle to dispose of the large amount of waste that's created by rapid urbanisation for many reasons:

1) Money — poorer countries often can't afford to dispose of waste safely, e.g. toxic waste has to be treated and this can be expensive. There are often more urgent problems to spend limited funds on, e.g. healthcare.
2) Infrastructure — poorer countries don't have the infrastructure needed, e.g. poor roads in squatter settlements mean waste disposal lorries can't get in to collect rubbish.
3) Scale — the problem is huge. A large city will generate thousands of tonnes of waste every day.

Air and Water Pollution Have Many Effects

Air pollution

Effects:

- Air pollution can lead to acid rain, which damages buildings and vegetation.
- It can cause health problems like headaches and bronchitis.
- Some pollutants destroy the ozone layer, which protects us from the sun's harmful rays.

Management of the pollution:

This can involve setting air quality standards for industries and constantly monitoring levels of pollutants to check they're safe.

Water pollution

Effects:

- Water pollution kills fish and other aquatic animals, which disrupts food chains.
- Harmful chemicals can build up in the food chain and poison humans who eat fish from the polluted water.
- Contamination of water supplies with sewage can spread diseases like typhoid.

Management of the pollution:

This can involve building sewage treatment plants and passing laws forcing factories to remove pollutants from their waste water.

Managing air and water pollution costs a lot of money and requires lots of different resources, e.g. skilled workers and good infrastructure. This makes it harder for poorer countries to manage pollution.

Waste disposal problems — what a load of rubbish...

The UK has laws that help to stop air and water pollution reaching dangerous levels, but many poorer countries have no regulations — you could pack a gas mask if you're visiting one, and a dry suit if you plan on having a dip...

Urbanisation — Case Study

China's population is 1.3 billion. That's 21 times as many as the UK or 400 million times as many as my house.

Push and Pull Factors have Caused Urbanisation in China

Urbanisation in China is being caused by the internal migration of people from rural areas to urban areas (rural-urban migration). In 1990 around 26% of the population lived in urban areas, but by 2006 a whopping 44% did (over 550 million people). Urbanisation is happening because of push and pull factors.

PUSH factors from rural areas:

1) Fewer jobs — more machinery has made farming more efficient, so fewer workers are needed. This has created high unemployment, e.g. 150 million rural people were unemployed in 2004.
2) Lower wages and higher poverty — wages are normally lower in rural areas in China, leading to more poverty. E.g. in 2004 there were 26.1 million people in rural areas in poverty.
3) Shortage of services — services like education and healthcare are funded by taxes collected within the local area. This means poor rural areas don't have the money to improve their services.

PULL factors to urban areas:

1) More jobs — there are more industries and jobs in urban areas.
2) Higher wages and lower poverty — the average income is three times higher in urban areas than in rural areas.
3) Better services — there are more (and better) education and healthcare services in urban areas because there's more money to pay for them. For example, lack of funding in rural areas meant that in 2002, 1.1 million children couldn't go to primary school.

Urbanisation in China has had Lots of Impacts

There are impacts of urbanisation in the urban areas, e.g. Beijing:

1) Positive — more workers and an increase in the demand for services in the urban areas helps to increase trade and industry. This is good for the economy.
2) Negative — the increasing population causes more pollution and environmental damage. Over 270 cities in China have no water treatment plants so sewage is dumped straight into local rivers.

And yes, there are impacts for the rural areas too:

1) Positive — about 130 million people who've left rural areas to work in towns send money home to their families. This increases their income and helps them to avoid poverty.
2) Negative — it's usually the young people who migrate, leaving an ageing population behind. About half of all Chinese people aged over 60 now live without any younger relatives to help look after them.

Urbanisation is Being Managed in China

China is trying to manage the problems of urbanisation in both rural and urban areas. For example:

1) Urban — in 2001, China changed its water supply system so it could cope with the increased sewage and pollution in urban areas. This helped to improve both water quality and supply.
2) Rural — in 2009, a pilot pension scheme was set up to give retired farmers a pension every month. This will help to raise income and reduce poverty in rural areas.

Excitement was in the air over the 2008 Olympics, and also lead and benzene...

Sounds like there's an awful lot of people rushing around China — that's one stampede I wouldn't want to get caught in. I can't cope with the stampede in the shopping centre at sales time. Just the thought makes me want to lie down.

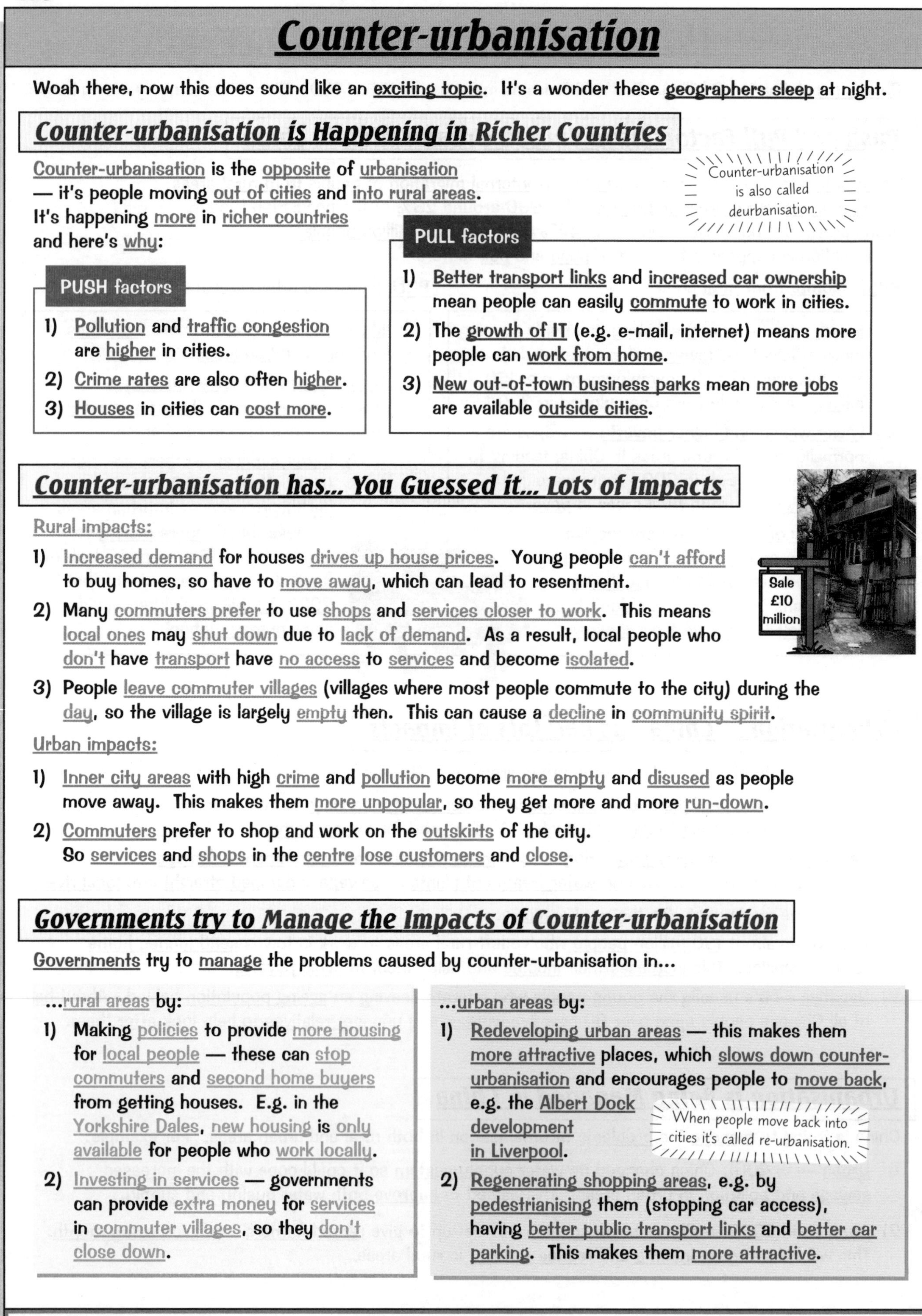

Counter-urbanisation

Woah there, now this does sound like an exciting topic. It's a wonder these geographers sleep at night.

Counter-urbanisation is Happening in Richer Countries

Counter-urbanisation is the opposite of urbanisation — it's people moving out of cities and into rural areas.
It's happening more in richer countries and here's why:

Counter-urbanisation is also called deurbanisation.

PUSH factors

1) Pollution and traffic congestion are higher in cities.
2) Crime rates are also often higher.
3) Houses in cities can cost more.

PULL factors

1) Better transport links and increased car ownership mean people can easily commute to work in cities.
2) The growth of IT (e.g. e-mail, internet) means more people can work from home.
3) New out-of-town business parks mean more jobs are available outside cities.

Counter-urbanisation has... You Guessed it... Lots of Impacts

Rural impacts:

1) Increased demand for houses drives up house prices. Young people can't afford to buy homes, so have to move away, which can lead to resentment.
2) Many commuters prefer to use shops and services closer to work. This means local ones may shut down due to lack of demand. As a result, local people who don't have transport have no access to services and become isolated.
3) People leave commuter villages (villages where most people commute to the city) during the day, so the village is largely empty then. This can cause a decline in community spirit.

Urban impacts:

1) Inner city areas with high crime and pollution become more empty and disused as people move away. This makes them more unpopular, so they get more and more run-down.
2) Commuters prefer to shop and work on the outskirts of the city. So services and shops in the centre lose customers and close.

Governments try to Manage the Impacts of Counter-urbanisation

Governments try to manage the problems caused by counter-urbanisation in...

...rural areas by:

1) Making policies to provide more housing for local people — these can stop commuters and second home buyers from getting houses. E.g. in the Yorkshire Dales, new housing is only available for people who work locally.
2) Investing in services — governments can provide extra money for services in commuter villages, so they don't close down.

...urban areas by:

1) Redeveloping urban areas — this makes them more attractive places, which slows down counter-urbanisation and encourages people to move back, e.g. the Albert Dock development in Liverpool.
2) Regenerating shopping areas, e.g. by pedestrianising them (stopping car access), having better public transport links and better car parking. This makes them more attractive.

When people move back into cities it's called re-urbanisation.

Right, that's it, everybody just STAY WHERE YOU ARE...

That's the trouble with society today — everyone's scurrying about, from town to country and back again, thinking the grass will be greener in the next place. Now if you'll excuse me, the removal van is here and I still have boxes to pack.

Sustainable Cities

Cripes, it's the 's' word again — it wouldn't be a geography section without it. This time we'll be discussing sustainable cities. Before starting, please put your eyelids in the open position and turn off your snoring...

Urban Areas Need to Become More Sustainable

1) Sustainable living means doing things in a way that lets the people living now have the things they need, but without reducing the ability of people in the future to meet their needs.
2) Basically, it means behaving in a way that doesn't irreversibly damage the environment or use up resources faster than they can be replaced.
3) For example, using only fossil fuels for power will add to climate change and eventually use them all up. This means the people in the future won't have any and the environment will be damaged — it's unsustainable.
4) Big cities need so many resources that it's unlikely they'd ever be truly sustainable. But things can be done to make a city (and the way people live there) more sustainable:

Schemes to reduce waste and safely dispose of it

More recycling means fewer resources are used, e.g. metal cans can be melted down and used to make more cans. Less waste is produced, which reduces the amount that goes to landfill. Landfill is unsustainable as it wastes resources that could be recycled and eventually there'll be nowhere left to bury the waste.

Safely disposing of toxic waste helps to prevent air and water pollution.

Conserving natural environments and historic buildings

Historic buildings, natural environments and open spaces are resources. If they get used up by people today (i.e. built on, or knocked down), they won't be available for people in the future to use. Historic buildings can be restored and natural environments can be protected. Existing areas of green space, like parks, should be left alone.

Building on brownfield sites

Brownfield sites are derelict areas that have been used, but aren't being used anymore. Using brownfield sites for new buildings stops green space being used up. So the space will still be available for people in the future. Developing brownfield sites also makes the city look nicer.

Building carbon-neutral homes

Carbon-neutral homes are buildings that generate as much energy as they use, e.g. by using solar panels to produce energy. For example, BedZED is a carbon-neutral housing development in London. More homes can be provided, without damaging the environment too much or causing much more pollution.

Creating an efficient public transport system

Good public transport systems mean fewer cars on the road, so pollution is reduced. Bus, train and tram systems that use less fuel and give out less pollution can also be used, e.g. some buses in London are powered by hydrogen and only emit water vapour.

5) People are much more likely to support sustainability initiatives like increased recycling or new public transport systems if they're involved in making the decisions about them. Including local people makes the schemes much more likely to succeed.

Conserve water — if it's brown, flush it down, if it's yellow, let it mellow...

Sustainability's a tough one to get your head around. Make sure you're clear on what it means before you memorise all the attempts to make it happen. And yes, toxic waste needs to be disposed of, not used to make fish with three eyes.

Sustainable Cities — Case Study

The key points to remember for this case study are: where the sustainable city is, its size, how the city is trying to be sustainable, what it costs, and how successful it is in being sustainable.

Curitiba is Aiming to be a Sustainable City

1) Curitiba is a city in southern Brazil with a population of 1.8 million people.
2) The overall aims of its planners are to improve the environment, reduce pollution and waste, and improve the quality of life of residents.
3) The city has a budget of $600 million to spend every year.
4) Curitiba is working towards sustainability in different ways:

1) Reducing car use
   - There's a good bus system, used by more than 1.4 million passengers per day.
   - It's an 'express' bus system — they have special pre-pay boarding stations that reduce boarding times, and bus-only lanes on the roads that speed up journeys.
   - The same cheap fare is paid for all journeys, which benefits poorer residents who tend to live on the outskirts of the city.
   - There are over 200 km of bike paths in the city.
   - The bus system and bike paths are so popular that car use is 25% lower than the national average and Curitiba has one of the lowest levels of air pollution in Brazil.

2) Plenty of open spaces and conserved natural environments
   - Green space increased from 0.5 m² per person in 1970 to 52 m² per person in 1990.
   - It has over 1000 parks and natural areas. Many of these were created in areas prone to flooding, so that the land is useful but no serious damage would be done if it flooded.
   - Residents have planted 1.5 million trees along the city's streets.
   - Builders in Curitiba are given tax breaks if their building projects include green space.

3) Good recycling schemes
   - 70% of rubbish is recycled. Paper recycling saves the equivalent of 1200 trees per day.
   - Residents in poorer areas where the streets are too narrow for a weekly rubbish collection are given food and bus tickets for bringing their recycling in to local collection centres.

Curitiba has been Very Successful in its Aim to be Sustainable

1) The reduction in car use means that there's less pollution and use of fossil fuels. This means the environment won't be damaged so much for people in the future.
2) Leaving green, open spaces and conserving the natural environment means that people in the future will still be able to use the open spaces.
3) The high level of recycling means that fewer resources are used and less waste has to go to landfill. This means more resources will be available in the future.
4) Curitiba is also a nice place to live — 99% of its residents said in a recent survey that they were happy with their town.

I'M FROM THE FUTURE. I LIKE TO USE OPEN SPACES.

Oh Curitiba, it makes me happy...

I wish I lived in Curitiba. No traffic jams, buses that run on time, plenty of places to ride my bike and lots of lovely green parks. They even have a flock of sheep that goes around the parks to eat the grass instead of using lawnmowers.

Urban Development

Urban development can't just happen any old how — it takes an awful lot of planning...

Planners Look at Social, Economic and Environmental Needs

Planners look at the needs of the population when designing new developments. For example:

Social needs

1) More housing — this can be built on old industrial sites (brownfield sites) near the city centre, or on the rural-urban fringe for commuters.
2) More room for social activities — e.g. parks replace brownfield sites in cities and places for activities like golf are set up in the rural-urban fringe.
3) Better transport systems and routes — more roads can be built in the rural-urban fringe to cope with increased traffic. Also, more and better public transport systems need to be planned.

Economic needs

More jobs — business parks and out-of-town shopping centres can be built on undeveloped land (greenfield sites) in the rural-urban fringe where land is cheap and it's easily accessible.

1) More waste disposal systems — landfill sites can be built on the rural-urban fringe to cope with the increase in waste.
2) More green spaces — derelict land in cities can be turned back into open spaces.

Urban Development Case Study — Glasgow, Scotland

1) Glasgow has a population of over 580 000 people. The decline in traditional industries, such as shipbuilding, has led to changes to Glasgow's social, economic and environmental needs. For example:
   - Some areas have poor quality, high-rise housing that was built in the 1960s and needs replacing.
   - More jobs are needed as there's high unemployment in Glasgow. In 2008, 7.1% of working age people in Glasgow were unemployed, compared to 4.9% for the whole of Scotland.
2) But recently the city has been benefiting from major redevelopments. For example:

   The Clyde Waterfront Regeneration Project is helping to change old inner city industrial land, e.g. land that had shipbuilding yards on it. The land will now be used for business, recreational and residential developments. Over 200 projects costing over £5 billion are ongoing including the Glasgow Science Centre, new offices, shops, parks and improved transport services (e.g. new buses).

3) The developments aim to meet the needs of the local population:
   - Social needs — the projects will improve transport and leisure facilities, and provide new homes.
   - Economic needs — new businesses are being attracted into the area, bringing jobs. E.g. over 50 000 new jobs are being created by the Waterfront Project.
   - Environmental needs — the areas will be more attractive and have more green spaces. E.g. the Waterfront Project will have various parks and natural areas.
4) The planners have also tried to make the developments sustainable. For example:
   - The Clyde Waterfront Regeneration Project has bus links, and walking and cycling routes to encourage lower car use. This means less pollution and fewer greenhouse gases will be emitted. Also, less fossil fuel will be used, which saves resources for future use.
   - The development is on derelict, brownfield sites, which saves land.
   - A lot of material from old buildings has been reused, so fewer resources have been used up.

It's bonnie on the clyde now...

Hope you enjoyed that trip to the distant North. Maybe you could send a postcard next time. I do love a good postcard, me. I'll daydream about postcards while you cover the page and jot down the finer points of this case study.

Retail Services

And now for a page about shopping — you get to think about retail therapy and call it revision. Bargain.

Start by Learning These Terms

1) There are two types of consumer goods (things people buy):
   - High order goods — these are goods that are only bought occasionally and are usually more expensive, e.g. clothes, furniture and cars. They're also called comparison goods.
   - Low order goods — these are goods that are bought frequently and are usually quite cheap, e.g. milk, bread and newspapers. They're also called convenience goods.
2) The threshold population — the minimum population needed to support a shop. Shops that sell high order goods have a high threshold population.
3) The sphere of influence — the area that people come from to visit a shop or an area. Shops that sell high order goods have a large sphere of influence because people will travel a long way occasionally to buy expensive items. People won't travel a long way to buy things they need regularly, so shops that sell low order goods will have a small sphere of influence. The distance people will travel for a particular good or service is called its range.

Go on... you know you want to...

Different Shopping Areas have Different Characteristics

Here are some characteristics of retail services in urban and rural areas:

	SHOPPING AREA	LOCATION	GOODS SOLD	THRESHOLD POPULATION	SPHERE OF INFLUENCE
URBAN	City centre	CBD	High order, e.g. clothes and jewellery.	High — because they sell high order goods and the rent is expensive.	Large — they attract people from a wide area.
URBAN	Out-of-town shopping centre	Rural-urban fringe	High order, e.g. clothes and hardware.	Medium — they sell high order goods but the cost of rent is lower.	Large — they attract people from a wide area.
URBAN	Shopping parades (short rows of shops)	Suburbs	High and low order, e.g. newspapers and clothes.	Medium — they sell a mixture of goods and the cost of rent is lower than in the city centre.	Medium — they attract people from the nearby area.
URBAN	Corner shops	Inner city	Low order, e.g. newspapers and bread.	Low — because they sell goods that are bought often and rent is cheap.	Small — they only attract local customers.
RURAL	Village shops	Villages	Low order, e.g. newspapers and bread.	Low — because they sell goods that are bought often and rent is cheap.	Small — they only attract local customers.

The size of a settlement will also affect what shops can locate there — the bigger a settlement is, the greater its population, so shops will have more potential customers. So the larger a settlement, the more likely it is to have shops selling high order goods.

Some Rural Shops Sell High Order Goods

Shops in rural areas sell mainly low order goods (see above), but some specialist shops that have a large sphere of influence can be found there. People are willing to travel far to buy specialist, rural goods, e.g. caravans or walking equipment (they have a large range).

Homework — go into town and buy yourself some shoes...

Alright, just kidding. Your homework is actually to go into town and buy me some shoes. I like five-inch heels with polka dots, size 11. Then when you come back, see how much of this page you remember. Good luck to you.

Changing Retail Services — Case Study

We all know fashions come and go — for example, corner shops are out and out-of-town shopping centres are in. If you're not sure you can keep up with it all, maybe this page will help you along.

Retail Services Change Over Time

There have been major changes in the way we shop in the UK in the last 100 years.
This is mainly due to two factors:

CHANGES TO TRANSPORT

Car ownership has increased so people can travel further for their shopping. This means there are fewer, smaller convenience stores in rural areas, but there are more out-of-town shopping centres. They're built out of town because land is cheaper, there's more available and it's accessible with on-site parking.

CHANGING MARKET FORCES

1) Changing market forces means changes in the supply and demand for goods and retail services.
2) Supply is how easy and cheap it is to get products.
3) Demand is what products people want and how much they are willing to pay for them.
4) Basically, people now want a larger range of goods at cheaper prices.
5) Smaller, specialist shops can't meet this demand, but larger chain stores and supermarkets can — they have lots of different products under one roof at much cheaper prices, so people shop there instead.

Social habits and work patterns have also changed — people have less time to shop for the things they really need (e.g. food) but want more leisure shopping time (e.g. to shop for clothes). This means it's convenient to use supermarkets, which stock all different types of food all together.

Retail Services have Changed in South Yorkshire

South Yorkshire has a mixture of rural and urban areas, e.g. Sheffield and part of the Peak District.
In recent years the provision of retail services in this area has changed:

1) In 1990, a large out-of-town shopping centre called Meadowhall was built near Sheffield. The centre has 280 shops, is easily accessible by car and has 12 000 free parking spaces. Around 800 000 shoppers visit the centre every week.
2) The number of shops in Sheffield city centre has declined. Some, such as House of Fraser, have moved to Meadowhall where the rent is cheaper and they can have more space. Some have closed down, possibly because shoppers are going to Meadowhall instead. Early estimates suggested a 15% trade loss from the city centre due to the building of Meadowhall.
3) There are also fewer shops and Post Offices® in the surrounding rural villages, such as Hope. This is because more people own cars and travel to urban areas to do their shopping.

Sheffield city centre is now fighting back by redeveloping itself, improving parking and using a 'City Watch' scheme to reduce crime. (It's reet nice — why not go and have a look for yourself.)

I love geography — it encourages you to revise shopping...

It's a funny thing — lots of people talk wistfully about the good old days when there were lots of small shops with real quality and proper service. But I've seen them snaffling the last few tins of luxury baked beans down at Asda.

Revision Summary for Section 9

Phew, there was an awful lot going on in that section. It's almost time to put your feet up and relax with a cup of tea, but first find out whether you've taken in all the details with these questions.

1) Is most urbanisation happening today in rich countries or poor countries?
2) What is rural-urban migration?
3) Give one cause of rural-urban migration in poorer countries.
4) Describe the land use of the following parts of a UK city:
   a) CBD, b) the rural-urban fringe, c) the suburbs.
5) Give two problems that cities in rich countries have to deal with.
6) Describe two solutions to one of these problems.
7) Give an example of a successful attempt to revitalise a CBD.
8) Name three problems caused by the increasing number of cars on the roads in richer countries.
9) Give one solution to reduce the number of cars in cities.
10) Give one reason why there is ethnic segregation in many urban areas.
11) Give one strategy used to support the multicultural mixture of urban areas.
12) What is a squatter settlement?
13) Describe what life is like for residents in a typical squatter settlement.
14) What is the informal sector of the economy?
15) a) Name a project that has improved a squatter settlement.
    b) Give one environmental improvement, one social improvement and one economic improvement that has happened. How successful has the project been?
16) Give one reason why waste disposal is such a problem in the cities of many poorer countries?
17) Give three effects of water pollution.
18) How can air pollution be managed?
19) a) Give an example of a country you have studied where urbanisation is taking place.
    b) Give two push factors and two pull factors that cause urbanisation in that country.
20) What is counter-urbanisation?
21) Give three impacts of counter-urbanisation in rural areas and describe how they can be managed.
22) What is meant by a sustainable city?
23) Explain how reducing waste can help a city to be more sustainable.
24) Explain how building on brownfield sites can help a city to become more sustainable.
25) a) Name a city and describe three ways in which the city is trying to become more sustainable.
    b) How successful do you think the city has been in its aim to be more sustainable? Give reasons for your answer.
26) List one social, one economic and one environmental need that planners consider when designing new developments.
27) a) Give one economic and one social aim of a development project you have studied.
    b) Describe how sustainable the development project is.
28) What are low order goods?
29) Describe the goods sold, threshold population, sphere of influence and accessibility for shops in the following areas: a) CBD, b) rural-urban fringe.
30) What are the two main factors that have changed the way we shop in the UK in the last 100 years?
31) Describe how retail services in a named area have changed and what effects this has had on the surrounding area.

Change in the Rural–Urban Fringe

Roll up, roll up, the revision circus is in town. Well, in the rural-urban fringe, anyway — the area right at the edge of a city, where there are both urban land uses (e.g. factories) and rural land uses (e.g. farming).

The Rural–Urban Fringe is a Popular Site for Development

As the population of an urban area increases the urban area gets bigger. This is called urban sprawl. One way it gets bigger is by development of the rural-urban fringe. Developments often include:

1) Out-of-town retail outlets.
2) Leisure facilities, e.g. golf courses, riding stables.
3) New transport links, e.g. new motorways connecting cities.
4) Housing — more housing is built in existing villages.

EXAMPLE: Golf courses, thousands of houses and the M5 motorway have been built in the rural area between Gloucester and Cheltenham.

The rural-urban fringe is popular for development because:

1) There's plenty of land available and it's cheaper than in urban areas. Some developments are huge so can only be built where there's lots of land, e.g. retail outlets need lots of space for car parking.
2) It's easy to reach from the urban areas, e.g. people can quickly drive out to retail outlets or golf courses and there's plenty of room to park.

Development of the Rural–Urban Fringe has Impacts

1) Traffic noise and pollution increase as there's more traffic.
2) People already living there may feel the extra housing and developments spoil the area.
3) Farmers may be forced to sell their land so it can be built on, meaning they can't earn a living.
4) Wildlife habitats are destroyed by building on them.

Some urban areas have 'greenbelts' around them though — a ring of land where development is restricted.

The Number and Size of Commuter Villages is Increasing

1) Some people live in villages and commute (travel) to work in urban areas. They choose to live there because it's a nicer environment and there's less crime, pollution and noise than in urban areas. Villages where there are a lot of commuters are called commuter villages.
2) Transport has become cheaper and faster in the 20th century. Road and rail links have improved too. This means more people can live further away from where they work and still commute to work easily. This has increased the number and size of commuter villages.
3) As a village becomes more popular it can cause property prices to increase. It can also cause an increase in traffic congestion.

Commuter villages are sometimes called suburbanised villages.

Growing Commuter Villages have Certain Characteristics

Lots of services, e.g. shops, schools and restaurants.

Lots of middle-aged couples with children, professionals and wealthy retired people who have moved there from the city as it's a nicer environment.

Lots of new detached houses, converted barns or cottages and expensive estates.

Good public transport links.

Some jobs, e.g. in local shops.

The soundtrack to the rural-urban fringe — a mix of house, garage and barn...

...a toe-tapping combination. But before you get carried away and start strutting your stuff, make sure you can reel off the impacts of rural-urban fringe development. Try and remember at least three, but four would be even better.

Change in Rural Areas — Case Study

Rural areas are changing, and it's not just that the country air is getting 'fresher' at muck-spreading time. Nope, there are economic and social changes too. Read on and you'll soon see exactly what I mean.

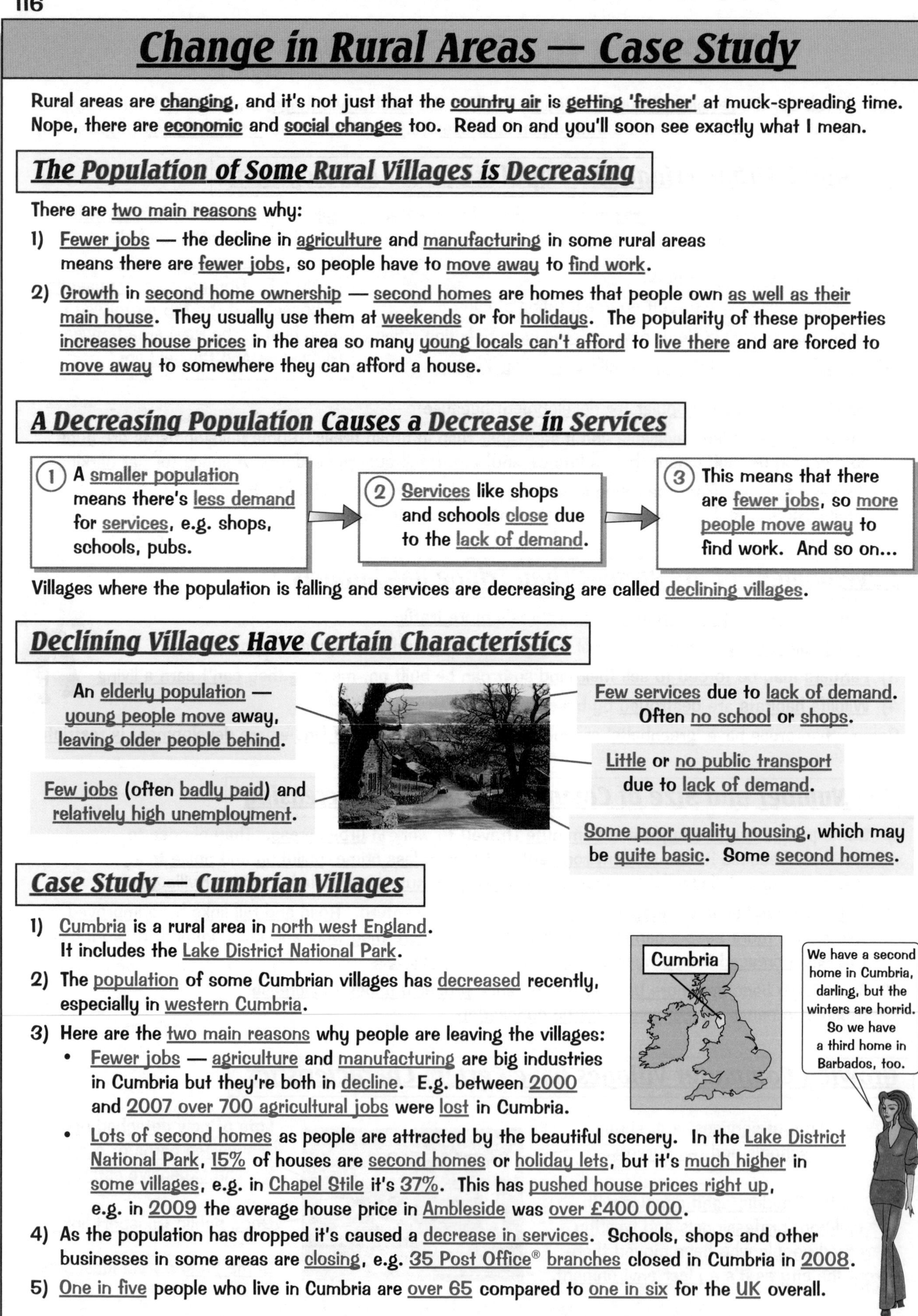

The Population of Some Rural Villages is Decreasing

There are two main reasons why:

1) Fewer jobs — the decline in agriculture and manufacturing in some rural areas means there are fewer jobs, so people have to move away to find work.
2) Growth in second home ownership — second homes are homes that people own as well as their main house. They usually use them at weekends or for holidays. The popularity of these properties increases house prices in the area so many young locals can't afford to live there and are forced to move away to somewhere they can afford a house.

A Decreasing Population Causes a Decrease in Services

1. A smaller population means there's less demand for services, e.g. shops, schools, pubs.
2. Services like shops and schools close due to the lack of demand.
3. This means that there are fewer jobs, so more people move away to find work. And so on...

Villages where the population is falling and services are decreasing are called declining villages.

Declining Villages Have Certain Characteristics

- An elderly population — young people move away, leaving older people behind.
- Few jobs (often badly paid) and relatively high unemployment.
- Few services due to lack of demand. Often no school or shops.
- Little or no public transport due to lack of demand.
- Some poor quality housing, which may be quite basic. Some second homes.

Case Study — Cumbrian Villages

1) Cumbria is a rural area in north west England. It includes the Lake District National Park.
2) The population of some Cumbrian villages has decreased recently, especially in western Cumbria.
3) Here are the two main reasons why people are leaving the villages:
   - Fewer jobs — agriculture and manufacturing are big industries in Cumbria but they're both in decline. E.g. between 2000 and 2007 over 700 agricultural jobs were lost in Cumbria.
   - Lots of second homes as people are attracted by the beautiful scenery. In the Lake District National Park, 15% of houses are second homes or holiday lets, but it's much higher in some villages, e.g. in Chapel Stile it's 37%. This has pushed house prices right up, e.g. in 2009 the average house price in Ambleside was over £400 000.
4) As the population has dropped it's caused a decrease in services. Schools, shops and other businesses in some areas are closing, e.g. 35 Post Office® branches closed in Cumbria in 2008.
5) One in five people who live in Cumbria are over 65 compared to one in six for the UK overall.

In my village it's cool to lie down — it's got a reclining population...

At this point I reckon you should check you know why people leave some rural villages, and why this makes it harder to buy a pint or post a parcel. Make sure you've a couple of case study facts stashed away too, to impress the examiner.

Change in UK Commercial Farming

Commercial farming (farming to make money) in some rural areas is changing. Farmer Giles has been thrown off his tiny tractor by Slick Rick and his giant combine harvester. At least that's how I understand it.

Agri-business has Replaced Traditional Farming in Some Areas

1) In the UK, 60 years ago, there used to be lots of small family farms that sold a mixture of produce. Now there are lots of large companies that own large farms. They often produce a single product.
2) This kind of large-scale commercial farming is called agri-business.
3) Modern farming practices used by agri-businesses help to maximise production and profits. But the practices can take their toll on the environment:

- Monoculture (growing just one type of crop) reduces biodiversity as there are fewer habitats.
- Removing hedgerows to increase the area of farmland destroys habitats. It also increases soil erosion (hedgerows normally act as windbreaks).
- Herbicides are used to maximise crop production, but they can kill wildflowers.
- Pesticides are used to maximise crop production, but they can kill other insects as well as pests.
- Fertilisers are also used to maximise crop production, but they can pollute rivers, killing fish.
- Making fertilisers, pesticides and herbicides uses fossil fuels, which adds to global warming.

Biodiversity is the number and variety of organisms. A habitat is where an organism lives.

When fertilisers pollute rivers it's called eutrophication.

Organic Farming is Also Becoming More Common

1) Organic farming is basically farming without using artificial pesticides or fertilisers.
2) Methods used include crop rotation (changing the type of crop that's planted every year to stop pests building up), using manure as a fertiliser, manual weeding and using biological control (e.g. using ladybirds instead of pesticides to kill aphids). These methods can be less damaging to the environment than other methods.
3) Organic farming is becoming more common, e.g. in the UK in 1998, 100 000 hectares of land were organically farmed and in 2003 this had increased to 700 000 hectares.
4) This is because the demand for organic food has increased. Some people buy organic food because they're concerned that modern farming practices damage the environment or that eating food that contains pesticide residues might be harmful.

Government Policies Aim to Reduce Farming's Environmental Impact

Here are two examples:

Environmental Stewardship Scheme

This involves paying farmers money for every hectare of land they manage in a way that reduces the environmental impact. For example, by farming organically.

Single Payment Scheme

This involves paying farmers a subsidy (paying them money to help them earn a decent living). But they're only paid it if they keep their land in a good environmental condition, e.g. if they leave 2 m around the edge of crop fields uncut to provide habitats. This encourages farmers to reduce the environmental impact of their farming.

Both schemes involve things like using fewer chemicals or leaving some areas uncultivated, so less food can be produced from the same area of land.
This can mean that more land has to be used for farming and the food produced is more expensive.

Monoculture is so 1990s — it's all about stereoculture these days...

It turns out farming isn't as simple as throwing a few seeds around a field and making the world's best scarecrow. Nope it's a serious business. Agri-business, in fact, so make sure you know the definition.

Change in UK Commercial Farming — Case Study

More exciting info on UK farming and a nice case study to get your teeth into. No need to thank me.

The Prices Farmers can Charge May be Decided by Supermarkets

1) Four major supermarket chains now control 75% of grocery sales in the UK.
2) This means farmers often have no choice but to cut their prices when asked to by the supermarkets, as there's no-one else to sell to. If they don't cut their prices the supermarkets will find other suppliers.
3) Many foods need processing before supermarkets will buy them, so sometimes farm products are bought by a processing firm. That firm then sells the finished product on to the supermarket at a profit. This adds another step to the supply chain, which can further reduce prices paid to farmers.
4) Some farmers struggle to earn a living due to the low prices (and some go out of business).

UK Farmers Now Have to Compete with the Global Market

1) Before the 1960s most of the food people ate was grown in the UK, usually in the local area.
2) Since then there's been an increase in the global trading of food with more and more of our food being imported from other countries.
3) This has helped to provide enough food for the growing population and has meant people in the UK can get a wide range of food all year round.
4) Imported food is often cheaper if it's grown in poorer countries where farmers pay less for land and pay less for workers to harvest it. UK farmers have to compete with these lower prices.

Transporting food a long way produces lots of CO_2, which adds to global warming.

Case Study — East Anglia

1) East Anglia is an area that includes Norfolk, Suffolk, Cambridgeshire and Essex.
2) It's known as the UK's 'bread basket' because it produces more than a quarter of England's wheat and barley. Farms in the area also produce 2.2 million eggs every day.
3) Agri-business has increased in East Anglia, e.g. in Essex the number of farms over 100 hectares increased from 828 in 1990 to 849 in 2005.

East Anglia

4) Organic farming in East Anglia has increased, but it's still quite low in the area, e.g. in 2008 1.3% of farmland was farmed organically, compared to 3.7% for England overall.
5) Farmers in East Anglia are trying to reduce the environmental impact of farming — the area has more land covered by the Environmental Stewardship Scheme (see previous page) than any other area in the UK.
6) Farmers in East Anglia have been affected by supermarket prices and competition from overseas. E.g. in 1997 peas from East Anglia sold at 25p per kilo but by 2002 this had dropped to 17p.

East Anglia — the home of bread and trigonometry...

Crikey, it sounds like being a farmer is a bit tougher nowadays. Make sure you know a case study for all things farming — examiners are absolutely baaaaaaaaaaaaaaaaaaaaaaarmy about them. Geddit? Ahem, maybe I should leave now...

Sustainable Rural Living

Ah, the countryside. It's all about running through the poppy fields flicking your shiny, newly-shampooed hair. It feels like it could go on forever. Perhaps it can, but only if we run and flick in a sustainable way.

Rural Living Needs to be Sustainable

Sustainable living means living in a way that lets the people alive now get the things they need, but without stopping people in the future getting the things they need. Basically it means behaving in a way that doesn't irreversibly damage the environment or use up resources faster than they can be replaced. There are two main reasons why rural living can be unsustainable:

1) High car use — many rural areas have little public transport (often due to low demand) so lots of people travel by car. This uses up fossil fuel resources and releases carbon dioxide, which adds to global warming.
2) Use of some farming techniques:
   - Some farming techniques use up fossil fuels. For example, some farms use a lot of artificial fertilisers — making artificial fertilisers uses up fossil fuel.
   - Irrigation of farmland can also deplete water resources.
   - Some techniques damage the environment (see page 117 for a list).

There are ways to make rural living more sustainable:

1) Conserve resources such as water and fossil fuels. For example, by:
   - Using public transport more to reduce car use.
   - Using irrigation techniques that don't waste water, e.g. drip-irrigation.
2) Protect the environment. For example, by:
   - Reducing the use of herbicides, fertilisers and pesticides to reduce their impacts.
   - Maintaining hedgerows to provide wildlife habitats.

Any attempts to make rural living sustainable have to support the needs of the rural population, e.g. you can't just stop all farming to reduce its environmental impacts because local people need a source of income.

Government Initiatives Protect the Rural Economy and Environment

There are various government initiatives (schemes) to help protect the rural environment and the rural economy — they help towards sustainable living. Here are a couple of examples:

1. Community Rail Partnerships help increase local train use by improving bus links, developing cycle routes to stations and improving station buildings. This reduces car use and the environmental impacts it has.

2. The Rural Development Programme for England gives farmers financial support to diversify their farms, e.g. to provide bed and breakfast accommodation or set up a tourist attraction. This gives farmers an extra income, so they're not as dependent on farming. It can also reduce the environmental impact of farming as some farmers don't need to farm as much.

3. The Environmental Stewardship Scheme involves paying farmers money to manage their land in a way that reduces the environmental impact (see page 117).

Rural living — it's not all tweed jackets, tractors and chewing straw...

First get your head around why rural living can be unsustainable, then even if you can't remember how to make it more sustainable in the exam, you should be able to work it out. Another pearl of wisdom... sometimes I think I'm too nice.

Changes to Farming in Tropical Areas

I love bananas in custard, but definitely not in crumble. This tropical farming page isn't about them though.

Subsistence Farming is being Replaced by Commercial Farming

1) Subsistence farming is where farmers only produce enough food to feed their families. In tropical areas farmers usually clear an area of rainforest to make land for producing food. The soil quickly becomes infertile though, so the farmers move to another area and start again. This is called shifting cultivation.
2) Subsistence farming is being replaced in some tropical areas by commercial farming — where crops and animals are produced to be sold, e.g. coffee, cotton, sugar cane and cattle.
3) Sometimes subsistence farmers switch to commercial farming, and sometimes big companies set up farms (they often take over subsistence farmers' land).

4) Commercial farm products are usually sold to richer countries.
5) This has a few impacts:
   - Subsistence farmers who've had their land taken by big companies are forced onto poorer land where it's harder to grow food for themselves.
   - If farmers are dependent on a single crop or animal and prices drop they might not have enough money to buy food. It also means farmers will only have an income around harvest or slaughter time — if they can't make a lot of money, they'll struggle to buy food for the rest of the year.
   - There isn't as much food being produced locally, so food has to be brought in from further away. This increases food prices.

Commercial farming in tropical areas can also be called cash cultivation.

Irrigation has Changed Agriculture

Growing lots of crops to sell needs a lot of water, so farmers often have to irrigate their land (artificially apply water). Irrigation has physical and human impacts:

	Positive	Negative
Physical impacts	• More land can be farmed. • Crop yields are higher and fewer harvests are lost due to lack of water. • High yields mean farmers don't need to clear more land for farming, e.g. by deforestation.	• Irrigation can cause soil erosion. • Without proper drainage salt can build up (salinisation) causing crops to fail. • If the land isn't well drained it becomes waterlogged so nothing can grow.
Human impacts	• Higher yields mean more food. This decreases the risk of famine. • Higher yields mean farmers make more profit — giving them a better quality of life.	• Large-scale irrigation projects can be expensive and cause rural debt to increase. • Mosquitoes that spread malaria breed in irrigation ditches. • Waterborne diseases can also become more common.

Appropriate Technologies have also Changed Agriculture

Appropriate technologies are simple, low cost technologies that increase food production. They're made and maintained using local knowledge and resources, so they're not dependent on any outside support, expensive equipment or fuel. Here are two examples:

1) The treadle pump is a human-powered pump used in Bangladesh. It pumps water from below the ground to irrigate small areas of land. This is important in Bangladesh as the main crop (rice) needs lots of water to grow. It costs US $7 to buy and it's increased Bangladeshi farmers' average annual incomes by roughly $100 because of increased crop yields.
2) Lines of stones are used to trap water on sloping fields in Burkina Faso. It increases the amount of water that soaks into the soil so more is available for crops. It's increased crop yields by about 50%.

The fanciest option isn't always the best, as many WAG weddings have proved...

The idea of appropriate technology is pretty important. You could be asked whether an irrigation system would be appropriate technology for a tropical region. Think about how it's suited to the area and what problems it could cause.

Factors Affecting Farming in Tropical Areas

Just a second, I hope I didn't just see your eyes glazing over. No, I didn't think I did. Don't worry, there are just a few more tropical farming issues to learn about and then this section is all finished.

Soil Erosion can be a Big Problem for Tropical Farmers

Soil erosion happens naturally due to the action of wind and rain. Soil erosion is common in tropical areas because there's heavy rainfall, which washes away the soil. Overgrazing can cause erosion because plants that hold the soil together are removed. Soil erosion can cause serious problems:

1) Erosion of the nutrient-rich top layer of soil makes the soil unsuitable for farming — it doesn't have enough nutrients and it can't hold water as well.
2) When the land can't be farmed anymore, the farmers either have to move away (e.g. to urban areas, see below) or they have to clear more land and start again.
3) The eroded soil is washed into rivers, which raises riverbeds. This means the rivers can't hold as much water and are more likely to flood.

Mining and Forestry Affect Subsistence Farming

A lot of mining and forestry goes on in tropical rainforests, which affects subsistence farming there:

MINING

1) Mining companies can force local people off their land. This means the local farmers have no source of income.
2) Mining uses lots of water. This can reduce crop yields for local farmers because there's less water for irrigation.
3) After the resources have all been extracted the land is often left unusable (e.g. because of pollution). This means there's less land available for local farmers.

FORESTRY

1) Deforestation can make floods more common as there are fewer trees to intercept rainfall. Floods can waterlog soil, reducing crop yields. They can also wash away crops.
2) Without trees, less water is removed from the soil and evaporated into the atmosphere. This means fewer clouds form and rainfall in the area is reduced. Reduced rainfall means lower crop yields for local farmers.
3) It's not all bad though — deforestation means more land is available for farming, so farmers can increase their income.

Farming Difficulties Lead to Rural-Urban Migration

1) Factors such as soil erosion, mining and forestry can cause farms to fail.
2) This means farmers can't make a profit or grow enough to feed themselves and their families.
3) People are forced to abandon their land and look for other work. They leave the countryside and move to towns and cities (this is known as rural-urban migration).
4) However, there aren't enough jobs or houses in the cities for all the people that move there. This means things like squatter settlements spring up (poor quality houses built illegally).
5) As more land is abandoned, less food is produced by the country. This causes food prices to rise due to the cost of importing it from other countries.
6) Governments can reduce rural-urban migration by helping farmers, e.g. by encouraging the use of appropriate technology to decrease water shortages and educating farmers about sustainable farming methods.

What a lot of cow pats — I blame failing farms...

More impacts to learn again here, but I guess you're getting used to that by now. If not, I recommend a smidge more revision. Cover the page and write down as many impacts of mining and forestry on subsistence farming as you can.

Revision Summary for Section 10

At last, the end of another long, hard section — well nearly the end. Before you stop for a well-earned cup of tea and an episode of Hollyoaks, there's just the small matter of this list of questions. It's really in your best interests to have a look through them, because if there are any you can't answer then you can bet your favourite pair of pants that that's exactly what will come up in the exam.

1) List four types of development often found in the rural-urban fringe.
2) Give two reasons why the rural-urban fringe is a popular place for these developments.
3) Give four impacts of development in the rural-urban fringe.
4) Give two reasons why people live in commuter villages.
5) List four characteristics of commuter villages.
6) What are the two main reasons why the populations of some rural villages are decreasing?
7) Explain why a decrease in population in a village can cause a decrease in services.
8) Give four characteristics of declining villages.
9) a) Give an example of a rural area in the UK with a declining population.
   b) Give one cause of depopulation in that area.
   c) Give one problem resulting from depopulation in that area.
10) Define the term agri-business.
11) What is monoculture?
12) Give two ways modern farming practices can affect the environment.
13) Explain what organic farming is.
14) Describe one government policy aimed at reducing the environmental impact of farming.
15) Explain how supermarkets may influence the prices farmers charge.
16) How does competition from the global market affect the prices UK farmers can get for their produce?
17) a) Give an example of a commercial farming area in the UK.
   b) Give two ways farming in the area has changed recently.
   c) Give one way the environmental impact of farming in the area has been reduced.
18) What is meant by sustainable living?
19) Give two ways that rural living can be unsustainable.
20) Describe two ways rural living can be made more sustainable.
21) Describe three government initiatives that are designed to boost the economy or protect the environment in rural areas.
22) Explain what is meant by:
   a) Subsistence farming.
   b) Commercial farming.
23) Give three problems switching from subsistence farming to commercial farming can cause.
24) Give two physical and two human impacts of irrigation.
25) What is meant by appropriate technology?
26) Explain why soil erosion is a problem for tropical farmers.
27) How can mining in an area have a negative impact on subsistence farming?
28) How can forestry in an area have a negative impact on subsistence farming?
29) Explain how factors such as soil erosion, mining and forestry can lead to rural-urban migration by farmers in tropical countries.

Measuring Development

OK, this topic's a bit trickier than the other human ones, but fear not — I'll take it slowly.

Development is when a Country is Improving

1) When a country develops it basically gets better for the people living there — their quality of life improves (e.g. their wealth, health and safety). Some people think quality of life just includes wealth, but it doesn't. (When you're just on about wealth it's usually referred to as economic development.)
2) The level of development is different in different countries, e.g. France is more developed than Ethiopia.
3) Development is pretty hard to measure because it` includes so many things. But you can compare the development of different countries using 'measures of development'. Here are a few:

NAME	WHAT IT IS	A MEASURE OF...	AS A COUNTRY DEVELOPS IT GETS...
Gross Domestic Product (GDP)	The total value of goods and services a country produces in a year. It's often given in US dollars (US$).	Wealth	Higher ↑
Gross National Income (GNI)	The total value of goods and services people of that nationality produce in a year (i.e. GDP + money from people living abroad). It's often given in US$. It's also called Gross National Product (GNP).	Wealth	Higher ↑
GNI per head	This is the GNI divided by the population of a country. It's sometimes called GNI per capita.	Wealth	Higher ↑
Birth rate	The number of live babies born per thousand of the population per year.	Womens' rights	Lower ↓
Death rate	The number of deaths per thousand of the population per year.	Health	Lower ↓
Infant mortality rate	The number of babies who die under 1 year old, per thousand babies born.	Health	Lower ↓
People per doctor	The average number of people for each doctor.	Health	Lower ↓
Literacy rate	The percentage of adults who can read and write.	Education	Higher ↑
Access to safe water	The percentage of people who can get clean drinking water.	Health	Higher ↑
Life expectancy	The average age a person can expect to live to.	Health	Higher ↑
Human Development Index (HDI)	This is a number that's calculated using life expectancy, literacy rate, education level (e.g. degree) and income per head.	Lots of things	Higher ↑

Many of these measures are linked — there's a relationship between them (the posh name for this is a correlation). For example, countries with high GNI tend to have low death rates and high life expectancy because they have more money to spend on healthcare.

These Measures Have Disadvantages

1) Economic measures can be inaccurate for countries where trade (the exchange of goods and services) is informal (not taxed). They're also affected by exchange rate changes (they're often given in US$).
2) The measures can be misleading when used on their own because they're averages — they don't show up elite groups in the population or variations within the country. For example, if you looked at the GNI of Iran it might seem quite developed (because the GNI is quite high), but in reality there are some really wealthy people and some poor people.
3) They also shouldn't be used on their own because as a country develops, some aspects develop before others. So it might seem that a country's more developed than it actually is.
4) Using more than one measure or using the HDI (which uses lots of measures) avoids these problems.

Measures of revision — they're called exams...

...and talking of exams you'd better get learnin' the 11 measures of development listed above. They could come up in the exam, so it'll help if you know what each one means and whether it gets higher or lower as a country develops.

Global Inequalities

'Global inequalities' means the level of development of different countries in the world is unequal.

Some Countries are More Developed than Others

1) Countries used to be classified into two categories based on how economically developed they were.
2) Richer countries were classed as More Economically Developed Countries (MEDCs) and poorer countries were classed as Less Economically Developed Countries (LEDCs).
3) MEDCs were generally found in the north. They included the USA, European countries, Australia and New Zealand.
4) LEDCs were generally found in the south. They included India, China, Mexico, Brazil and all the African countries.
5) But using this simple classification you couldn't tell which countries were developing quickly and which weren't really developing at all. Nowadays, countries are classified into more categories, for example:

Rich industrial countries — these are the most developed countries in the world. For example, the UK, Norway, USA, Canada, France.

Former communist countries — these countries aren't really poor, but aren't rich either (they're kind of in the middle). They're developing quickly, but not as quick as NICs are. For example, the Czech Republic, Bulgaria, Poland.

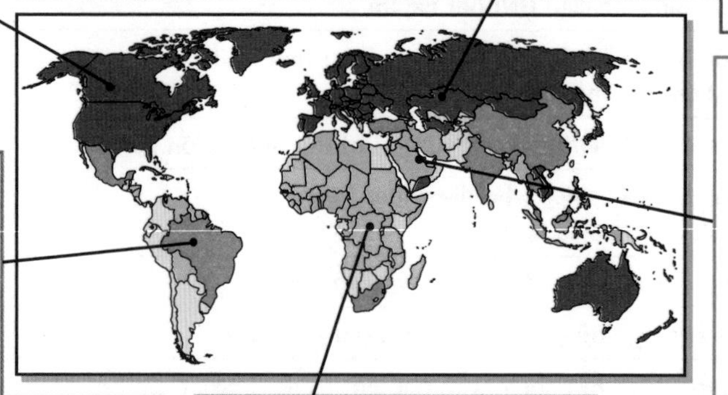

Oil-exporting countries — these are quite rich (they have a high GNI) but the wealth often belongs to a few people and the rest are quite poor. For example, Qatar, Kuwait, Saudi Arabia.

Newly Industrialising Countries (NICs) — these are rapidly getting richer as their economy is moving from being based on primary industry (e.g. agriculture) to secondary industry (manufacturing). For example, China, India, Brazil, Mexico, South Africa, Taiwan.

Heavily indebted poor countries — these are the poorest, least developed countries in the world. For example, Ethiopia, Chad, Angola.

Quality of Life Isn't the Same as Standard of Living

1) As a country develops the quality of life and standard of living of the people who live there improves.
2) Someone's standard of living is their material wealth, e.g. their income, whether they own a car.
3) Quality of life includes standard of living and other things that aren't easy to measure, e.g. how safe they are and how nice their environment is.
4) In general, the higher a person's standard of living the higher their quality of life. But just because they have a high standard of living doesn't mean they have a good quality of life. For example, a person might earn loads and have a flash car, but live somewhere where there's lots of crime and pollution.
5) Different people in different parts of the world have different ideas about what an acceptable quality of life is. For example, people in the UK might think it means having a nice house, owning a car, and having access to leisure facilities — people in Ethiopia might think it means having clean drinking water, plenty of food, somewhere to live and no threat of violence.

My quality of life would be improved by more cake...

Unfortunately, revising development categories is a teeny bit more difficult now that there are a few categories rather than just two (I wish those geographers would make their minds up about what to call things — sooooo inconsiderate).

Causes of Global Inequalities

Zzzzzz... oh, sorry, I nodded off for a minute there. There are plenty of reasons why global inequalities exist — i.e. why countries differ in how developed they are. Better get cracking then...

Environmental Factors Affect How Developed a Country Is

A country is more likely to be less developed if it has...

1 A POOR CLIMATE

1) If a country has a poor climate (really hot or really cold) they won't be able to grow much.
2) This reduces the amount of food produced.
3) In some countries this can lead to malnutrition, e.g. in Chad and Ethiopia. People who are malnourished have a low quality of life.
4) People also have fewer crops to sell, so less money to spend on goods and services. This also reduces their quality of life.
5) The government gets less money from taxes (as less is sold and bought). This means there's less to spend on developing the country, e.g. to spend on improving healthcare and education.

2 POOR FARMING LAND

If the land in a country is steep or has poor soil (or no soil) then they won't produce a lot of food. This has the same effect as a poor climate (see above).

3 LIMITED WATER SUPPLIES

Some countries don't have a lot of water, e.g. Egypt, Jordan. This makes it harder for them to produce a lot of food. This has the same effect as a poor climate (see above).

4 LOTS of NATURAL HAZARDS

1) A natural hazard is an event that has the potential to affect people's lives or property, e.g. earthquakes, tsunamis, volcanic eruptions, tropical storms, droughts, floods.
2) When natural hazards do affect people's lives or property they're called natural disasters.
3) Countries that have a lot of natural disasters have to spend a lot of money rebuilding after disasters occur, e.g. Bangladesh.
4) So natural disasters reduce quality of life for the people affected, and they reduce the amount of money the government has to spend on development projects.

5 FEW RAW MATERIALS

1) Countries without many raw materials like coal, oil or metal ores tend to make less money because they've got fewer products to sell.
2) This means they have less money to spend on development.
3) Some countries do have a lot of raw materials but still aren't very developed because they don't have the money to develop the infrastructure to exploit them (e.g. roads and ports).

There are Three Main Political Factors that Slow Development

1) If a country has an unstable government it might not invest in things like healthcare, education and improving the economy. This leads to slow development (or no development at all).
2) Some governments are corrupt. This means that some people in the country get richer (by breaking the law) while the others stay poor and have a low quality of life.
3) If there's war in a country the country loses money that could be spent on development — equipment is expensive, buildings get destroyed and fewer people work (because they're fighting). War also directly reduces the quality of life of the people in the country.

Hot and dry — good for holidays, bad for development...

Basically, if a country is rubbish for farming, tropical storms keep wrecking the place or it has a dodgy government, then it's going to struggle to develop. There are a few exceptions though, e.g. Japan gets battered by natural hazards but is developed.

Causes of Global Inequalities

Countries really do have a tough time trying to develop. It's not just things like earthquakes and a shortage of water that hold them back — things like trade problems, debt and mucky water are to blame too...

Economic Factors Affecting Development Include Trade and Debt

A country is more likely to be less developed if it has...

1 POOR TRADE LINKS

1) Trade is the exchange of goods and services between countries.
2) World trade patterns (who trades with who) seriously influence a country's economy and so affect their level of development.
3) If a country has poor trade links (it trades a small amount with only a few countries) it won't make a lot of money, so there'll be less to spend on development.

2 LOTS of DEBT

1) Very poor countries borrow money from other countries and international organisations, e.g. to help cope with the aftermath of a natural disaster.
2) This money has to be paid back (sometimes with interest).
3) Any money a country makes is used to pay back the money, so isn't used to develop.

3 AN ECONOMY BASED ON PRIMARY PRODUCTS

1) Countries that mostly export primary products (raw materials like wood, metal and stone) tend to be less developed.
2) This is because you don't make much profit by selling primary products. Their prices also fluctuate — sometimes the price falls below the cost of production.
3) This means people don't make much money, so the government has less to spend on development.
4) Countries that export manufactured goods tend to be more developed.
5) This is because you usually make a decent profit by selling manufactured goods. Wealthy countries can also force down the price of raw materials that they buy from poorer countries.

Social Factors Affect Development Too

1 DRINKING WATER

1) A country will be more developed if it has clean drinking water available.
2) If the only water people can drink is dirty then they'll get ill — waterborne diseases include typhoid and cholera. Being ill a lot reduces a person's quality of life.
3) Ill people can't work, so they don't add money to the economy, and they also cost money to treat.
4) So if a country has unsafe drinking water they'll have more ill people and so less money to develop.

2 THE PLACE OF WOMEN IN SOCIETY

1) A country will be more developed if women have an equal place with men in society.
2) Women who have an equal place in society are more likely to be educated and to work.
3) Women who are educated and work have a better quality of life, and the country has more money to spend on development because there are more people contributing to the economy.

3 CHILD EDUCATION

1) The more children that go to school (rather than work) the more developed a country will be.
2) This is because they'll get a better education and so will get better jobs. Being educated and having a good job improves the person's quality of life and increases the money the country has to spend on development.

Debt — the cause of slow development and my lack of a fancy mobile phone...

Reading this page makes you realise how nice the UK is — there's trade galore, clean water and women work. Don't forget there are always exceptions to the rules above, e.g. countries that export oil (a primary product) are often quite rich.

Global Inequalities — Case Study

Thought you could get away without learning a case study for why there are global inequalities (why some countries are less developed than others), think again buster...

Hurricane Mitch Hit Nicaragua and Honduras in October 1998

Hurricane Mitch hit a few countries in 1998, but Nicaragua and Honduras were the worst hit. Here are some of the impacts in each country:

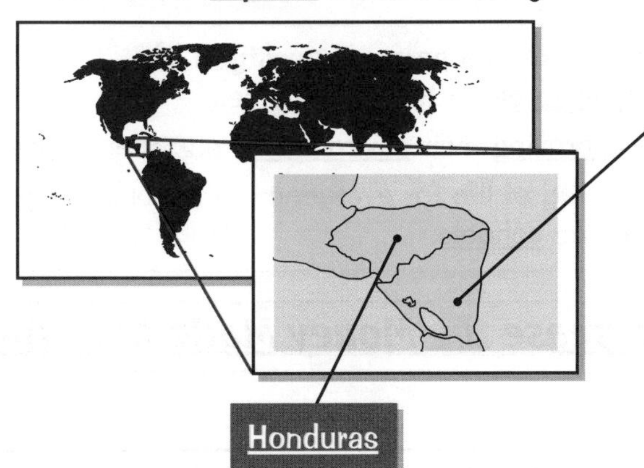

Nicaragua

1) Around 3000 people were killed.
2) The impact on agriculture was high — crops failed and 50 000 animals died.
3) 70% of roads were unusable and 71 bridges were damaged or destroyed.
4) 23 900 houses were destroyed and 17 600 more were damaged.
5) 340 schools and 90 health centres were damaged or destroyed.

Honduras

1) Around 7000 people were killed.
2) The hurricane destroyed 70% of the country's crops, e.g. bananas, rice, coffee beans.
3) Around 70-80% of the transport infrastructure (e.g. roads and bridges) was severely damaged.
4) 35 000 houses were destroyed and 50 000 more were damaged.
5) 20% of schools were damaged, as well as 117 health centres and six hospitals.

Hurricane Mitch Set Back Development in Nicaragua...

1) In 1998 the GDP grew by 4%, which was less than estimated. The rate of growth slowed in the later months of 1998 — after Hurricane Mitch hit.
2) Exports of rice and corn went down because crops were damaged by the hurricane. This meant people earnt less money, so were poorer, and the government had less to spend on development.
3) The total damage caused by the hurricane is estimated to be $1.2 billion. The cost of repairs took money away from development.
4) The education of children suffered — the number of children that worked (rather than went to school) increased by 8.1% after the hurricane. This meant the children had a lower quality of life and found it harder to get good jobs later in life.

...and In Honduras

1) In 1998 money from agriculture made up 27% of the country's GDP. In 2000 this had fallen to 18% because of the damage to crops caused by the hurricane. This reduced the quality of life for people who worked in agriculture because they made less money.
2) GDP was estimated to grow 5% in 1998, but it only grew 3% due to the hurricane. This meant there was less money available for development than there would have been if Mitch hadn't hit.
3) The cost of repairing and rebuilding houses, schools and hospitals was estimated to be $439 million — this money could have been used to develop the country.
4) All these things set back development — the Honduran President claimed the hurricane destroyed 50 years of progress.

Nicaragua is an anagram of iguana car — brum brum...

There are tons of facts on this page for you to cram into your brain. If you already know a case study of how a natural disaster affected development that's fine, but if not, shut the book and scribble what you can remember till you get it all.

Reducing Global Inequality

As you might have gathered, global inequality is on the list of bad things in the world (along with revision, exams and broccoli). So some people are trying to reduce it by helping poorer countries develop.

Some People are Trying to Improve Their Own Quality of Life

Some people in poorer countries try to improve their quality of life on their own — rather than relying on help from others. This is called 'self-help'. Here are a few ways that people do this:

1) Moving from rural areas to urban areas often improves a person's quality of life. Things like water, food and jobs are often easier to get in towns and cities.
2) Some people improve their quality of life by improving their environment, e.g. their houses.
3) Communities can work together to improve quality of life for everyone in the community, e.g. some communities build and run services like schools.

Fair Trade and Trading Groups Help Increase the Money Made from Trade

Fair trade

1) Fair trade is all about getting a fair price for goods produced in poorer countries, e.g. coffee.
2) Companies who want to sell products labelled as 'fair trade' have to pay producers a fair price.
3) Buyers also pay extra on top of the fair price to help develop the area where the goods come from, e.g. to build schools or health centres.
4) Only producers that treat their employees well can take part in the scheme. E.g. producers aren't allowed to discriminate based on sex or race, and employees must have a safe working environment. This improves quality of life for the employees.
5) However, producers in a fair trade scheme often produce a lot because of the good prices — this can cause them to produce too much. An excess will make world prices fall and cause producers who aren't in a fair trade scheme to lose out.

A 'fair price' is a price that's high enough for the producer to make a profit.

Trading groups

1) These are groups of countries that make agreements to reduce barriers to trade (e.g. to reduce import taxes) — this increases trade between members of the group.
2) When a poor country joins a trading group, the amount of money the country gets from trading increases — more money means that more development can take place.
3) However, it's not easy for poorer countries that aren't part of trading groups to export goods to countries that are part of trading groups. This reduces the export income of non-trading group countries and slows down their development.

E.g. NAFTA is a trade group including the USA, Canada and Mexico.

The Debt of Poorer Countries can be Reduced

1) Debt abolition is when some or all of a country's debt is cancelled. This means they can use the money they make to develop rather than to pay back the debt. For example, Zambia (in southern Africa) had $4 billion of debt cancelled in 2005. In 2006, the country had enough money to start a free healthcare scheme for millions of people living in rural areas, which improved their quality of life.
2) Conservation swaps (debt-for-nature swaps) are when part of a country's debt is paid off by someone else in exchange for investment in conservation. For example, in 2008 the USA reduced Peru's debt by $25 million in exchange for conserving its rainforests.

Fair trade — weeks of revision and stress for a piece of paper with a grade on it...

Don't be glum, you're over halfway through this gem of a section now. There's just a little bit more revision to go until you're a fully-fledged development guru — something I know you've wanted to be since you were little...

Reducing Global Inequality

One way less developed countries are given a helping hand is through international aid.

Some Types of International Aid Speed Up Development

1) Aid is given by one country to another country in the form of money or resources (e.g. food, doctors).
2) The country that gives the aid is called the donor — the one that gets the aid is called the recipient.
3) There are two main sources of aid from donor countries — governments (paid for by taxes) and Non-Governmental Organisations (NGOs, paid for by voluntary donations).
4) There are two different ways donor governments can give aid to recipient countries:
   - Directly to the recipient — this is called bilateral aid.
   - Indirectly through an international organisation that distributes the aid — this is called multilateral aid.

International organisations include the United Nations (UN) and the World Bank.

5) Bilateral aid can be tied — this means it's given with the condition that the recipient country has to buy the goods and services it needs from the donor country. This helps the economy of the donor country. However, if the goods and services are expensive in the donor country, the aid doesn't go as far as it would if the goods and services were bought elsewhere.
6) Aid can be classed as either short-term or long-term depending on what it's used for:

SHORT-TERM AID

1) This is money or resources that help recipient countries cope during emergencies, e.g. floods.
2) The aid has an immediate impact so more people will survive the emergency.
3) There are disadvantages though:
   - The stage of development of the recipient country remains unchanged overall.
   - If either country is slow to react, aid may not get to where it's most needed.
   - The aid may not reach those who need it because of things like theft and transport problems.

LONG-TERM AID

1) This is money or resources that help recipient countries to develop, e.g:
   - It's used to build dams and wells to improve clean water supplies.
   - It's used to construct schools to improve literacy rates.
2) Over time, recipient countries become less reliant on foreign aid as they become more developed.
3) However, it can take a while before the aid benefits a country, e.g. hospitals take a long time to build.

7) For both types of aid the recipient may become dependent on the aid — they don't bother spending their own money developing themselves because they get it from someone else.
8) In some recipient countries aid is misused because they have corrupt governments — the government uses the money and resources to fund their lifestyle or to pay for political events.

Still or sparkling, minister?

International Aid may not be Sustainable

1) To be sustainable, aid must help development in ways that don't irreversibly damage the environment or use up resources (including money) faster than they can be replaced.
2) An example of a sustainable aid project would be a scheme that helps people switch from earning money by deforestation to earning money in a more environmentally friendly way. This reduces environmental damage and makes sure trees are still there for future generations.
3) An example of an unsustainable aid project would be investment in large, shallow water wells in areas with little rainfall. Use of the wells could use up water faster than it's replaced. This would mean that the amount of water available for future use would be reduced.

Watching Live Aid on DVD doesn't count as revision...

Quite a bit to remember on this page — try drawing a simple diagram to show all the different ways that aid goes from a donor country to a recipient country. If you're still a bit confused after that just give Bob Geldof a call...

Reducing Global Inequality — Case Study

There are loads of development projects going on around the world to reduce inequality.
Much as I'd like to, I can't possibly tell you about all of them, so here's an example of one...

FARM-Africa helps the Development of Rural Africa

1) FARM-Africa is a non-governmental organisation (NGO) that provides aid to eastern Africa.
2) It's funded by voluntary donations.
3) It was founded in 1985 to reduce rural poverty.
4) FARM-Africa runs programmes in five African countries — Ethiopia, Sudan, Kenya, Uganda and Tanzania.
5) FARM-Africa has been operating in Ethiopia since 1988. Here are four of the projects it runs there:

Project	Region	Problem	What's being done	Helping...	Sustainability
Rural Women's Empowerment	Various	There are very few opportunities for Ethiopian women to make money. This means they have a low quality of life and struggle to afford things like healthcare.	Women are given training and livestock to start farming. Loan schemes have been set up to help women launch small businesses like bakeries and coffee shops. Women have been given legal training to advise other women of their rights.	Around 15 160 people.	Once the new businesses have been set up they'll continue to grow and make money. This means that money will be available as a future resource.
Prosopis Management	Afar	Prosopis, a plant introduced by the government to stabilise soils, has become a pest — it invades grazing land, making farming difficult.	Farmers are shown how to convert prosopis into animal feed. The animal feed is then sold, generating a new source of income.	Around 4400 households.	Once the farmers have been taught the new technique they'll be able to carry on using it. This means that money will be available as a future resource.
Community Development Project	Semu Robi	Frequent droughts make farming very difficult. This reduces the farmer's income and can lead to malnutrition. Semu Robi is a remote region, so getting veterinary care for livestock is difficult.	People are given loans to buy small water pumps to irrigate their farmland. This reduces the effects of drought. People are trained in basic veterinary care so they can help keep livestock healthy.	Around 4100 people.	The project means people are able to farm more crops and animals. This means they can earn more money. But if too much water is used there won't be any left for other people.
Sustainable Forest Management	Bale	Forests are cut down to make land for growing crops and grazing livestock. Trees are also cut down for firewood. This reduces resources for future generations.	Communities are taught how to produce honey and grow wild coffee. These are then sold, so people can make money without cutting down trees. Communities are also taught how to make fuel-efficient stoves that use less wood. This also reduces deforestation.	Around 7500 communities.	Less deforestation means there'll still be trees for future generations. Also, people can make money themselves by selling the coffee and honey.

You'd need a seriously meaty plough to farm the whole of Africa...

Yes, it's another case study for you. I think this one's pretty interesting though, and it's not too difficult to get your head around. The more facts and figures you can remember in the exam, the more impressed the examiner will be...

Development Levels in the EU

Different countries in the EU have very different levels of development. And here's a lovely example...

Bulgaria is Less Developed than The UK

Bulgaria joined the EU in 2007 — it's less developed than the UK. For example:

- In 2007 Bulgaria had a GNI per head of $11 180 and the UK had a GNI per head of $33 800.
- Life expectancy in Bulgaria is six years lower (73 compared to 79).
- The HDI for Bulgaria is 0.824, whereas it's 0.947 for the UK.

Here are a few reasons why Bulgaria is less developed than the UK:

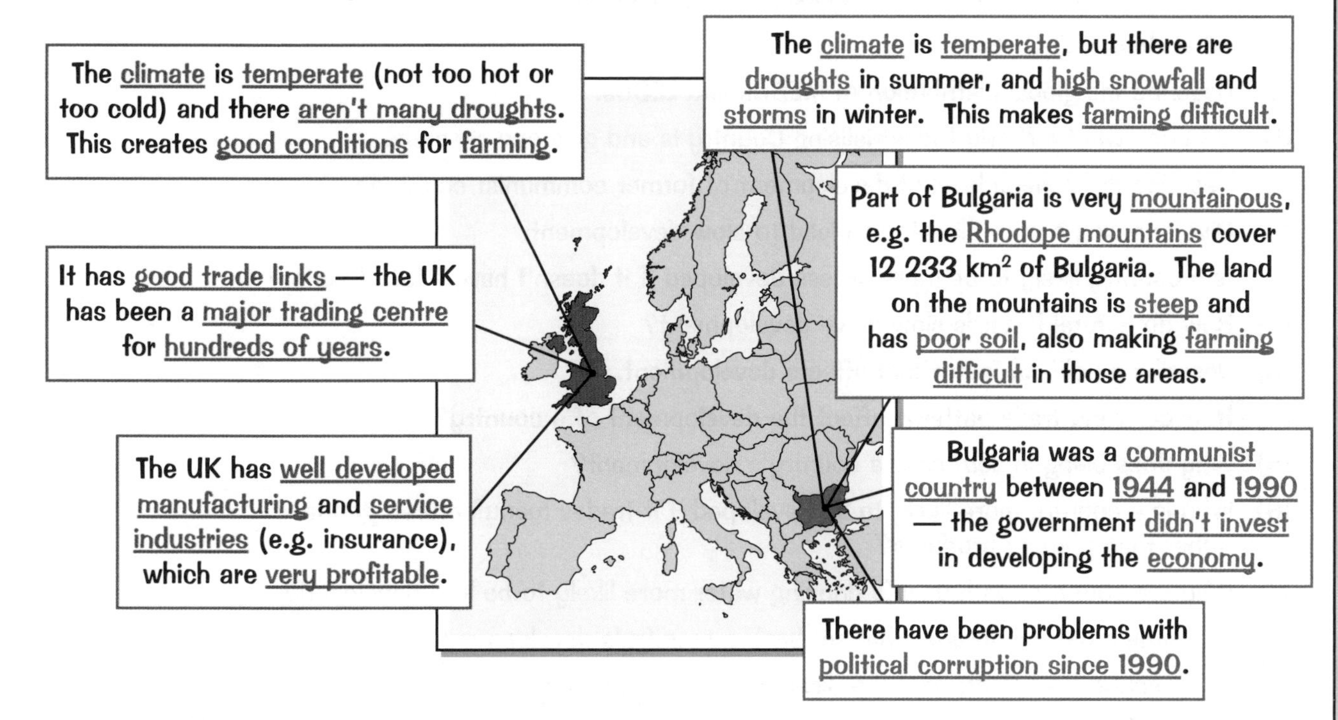

The EU is Trying to Reduce Inequalities

Here are some of the ways that the EU is trying to reduce inequality in and between its member countries:

1) The URBAN Community Initiative — money is given to certain EU cities to create jobs, reduce crime and increase the area of green space (e.g. parks).
2) The Common Agricultural Policy (CAP) — farmers are subsidised (paid) to grow certain products. Also, when world food prices are low, the EU buys produce and guarantees farmers a reasonable income. The CAP also puts a high import tax on foreign produce so people in the EU are more likely to buy food produced in the EU. All these things improve the quality of life for farmers.
3) Structural Funds — these provide money for research and development, improving employment opportunities, reducing discrimination and improving transport links. The aim of the fund is to get all members of the EU to a similar level of development (reducing inequalities within Europe).

The EU is also trying to develop Bulgaria in some specific ways:

- SAPARD (Special Accession Programme for Agriculture and Rural Development) gives money to Bulgaria and two other countries to invest in agriculture.
- Funds earmarked for Bulgaria have been partially frozen until the government shows it's making progress in fighting corruption.

Bulgaria's not as developed as the UK but it produces champion sumo wrestlers...

...seriously, I'm not joking, who would've thought it. Well, you've nearly made it to the end of this section, so you deserve a mini treat before you have a go at the revision summary — go and get a biscuit and a cup of tea.

Revision Summary for Section 11

Time to find out if you've developed a memory of development facts. Try these questions and if there are any that you don't know immediately have a look back at the page to refresh that big grey blob of yours. I know it's hard work, but it'll be worth it come the exam...

1) What is Gross National Income (GNI)?
2) Define birth rate.
3) Does literacy rate get higher or lower as a country develops?
4) Give three things that are used to calculate the HDI for a country.
5) Give one disadvantage of using measures of development.
6) What does MEDC stand for?
7) Describe the global distribution of MEDCs and LEDCs.
8) Describe what a Newly Industrialising Country is and give one example.
9) Describe the general level of development of former communist countries.
10) Give one way a poor climate can lead to slow development.
11) Is a country likely to be more or less developed if it doesn't have a lot of water?
12) How do natural hazards slow down development?
13) Describe a political factor that affects development.
14) How do world trade patterns affect the development of a country?
15) Why does being in debt slow a country's development?
16) Why is a country more likely to be developed if it trades manufactured goods rather than primary products?
17) Why are countries with unsafe drinking water more likely to be less developed?
18) a) Name a natural disaster that set back a country's development.
    b) Describe the effects that the disaster had on development.
19) Give two ways that people in poor countries are trying to improve their own quality of life.
20) What is fair trade?
21) How do trading groups help to reduce inequalities?
22) Give two ways the debt of a poor country can be reduced.
23) What is multilateral aid?
24) a) What is tied aid?
    b) Give one disadvantage of tied aid.
25) What's the difference between short-term and long-term aid?
26) Give one advantage of short-term aid.
27) a) Name a development project in a poorer country that you have studied.
    b) Describe how the development project is improving quality of life for the people in the area.
    c) Comment on how sustainable the development project is.
28) a) Name two countries in the EU that have contrasting levels of development.
    b) State three reasons why they have different levels of development.
    c) Give two ways in which the EU is trying to reduce the differences.

Types of Industry and Employment Structure

Right, it's the start of a brand new section. No need to go dashing ahead with excitement though — get your head around these basics first. Otherwise, you might find things get tricky later in the section.

There are Four Different Types of Industry

The four types are — primary, secondary, tertiary and quaternary. The employment structure of a country describes what proportion of its workforce is employed in each type of industry.

1) Primary industry involves collecting raw materials, e.g. farming, fishing, mining and forestry.
2) Secondary industry involves turning a product into another product (manufacturing), e.g. making textiles, furniture, chemicals, steel and cars.
3) Tertiary industry involves providing a service — anything from financial services, nursing and retail to the police force and transport.
4) Quaternary industry is high technology — where scientists and researchers investigate and develop new products, e.g. in the electronics and IT industry.

Quaternary industry is sometimes thought of as a part of tertiary industry.

A Country's Employment Structure Changes as it Develops

Less developed countries

1) Most of the workforce is employed in primary industry.
2) Few people work in secondary industry because there's not enough money to invest in the technology needed for this type of industry, e.g. to build large factories.
3) A small percentage of people work in tertiary industry — usually in cities where there are banks, hospitals and schools.
4) There's no quaternary industry because the country doesn't have enough educated or skilled workers, and it can't afford to invest in the technology needed, e.g telescopes.

Many workers in less developed countries don't appear in official statistics because they work in jobs that aren't taxed or monitored by the government, e.g. street traders. These jobs are referred to as the informal sector of the economy.

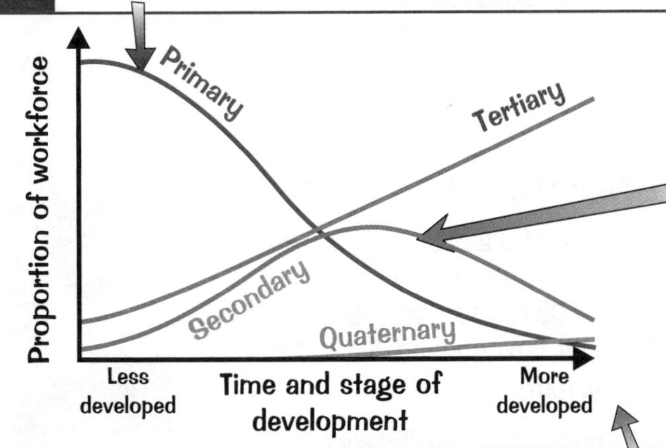

NICs are countries where the employment in secondary industry is increasing. As a country becomes more developed the percentage of people employed in secondary industry increases then decreases. This is because as infrastructure develops, businesses move their factories to less developed countries where labour is cheaper.

More developed countries

1) Few people work in primary industry because machines replace workers, and it's cheaper to import primary products from other countries, e.g. farm workers in poorer countries earn very little so the produce is cheap.
2) Fewer people work in secondary industry than in NICs (see above).
3) Most people work in tertiary industry because there's a skilled and educated workforce, and there's a high demand for services like banks and shops.
4) There's some quaternary industry because the country has lots of highly skilled labour and has money to invest in the technology needed.

Employment structure changes over time because countries become more wealthy, and education and infrastructure improve.

Primary industry — making small chairs, glitter and crayons...

There's a close link between a country's wealth and its employment structure. Primary industry doesn't make much profit whereas quaternary industry can make a lot. And more profit means a more developed country.

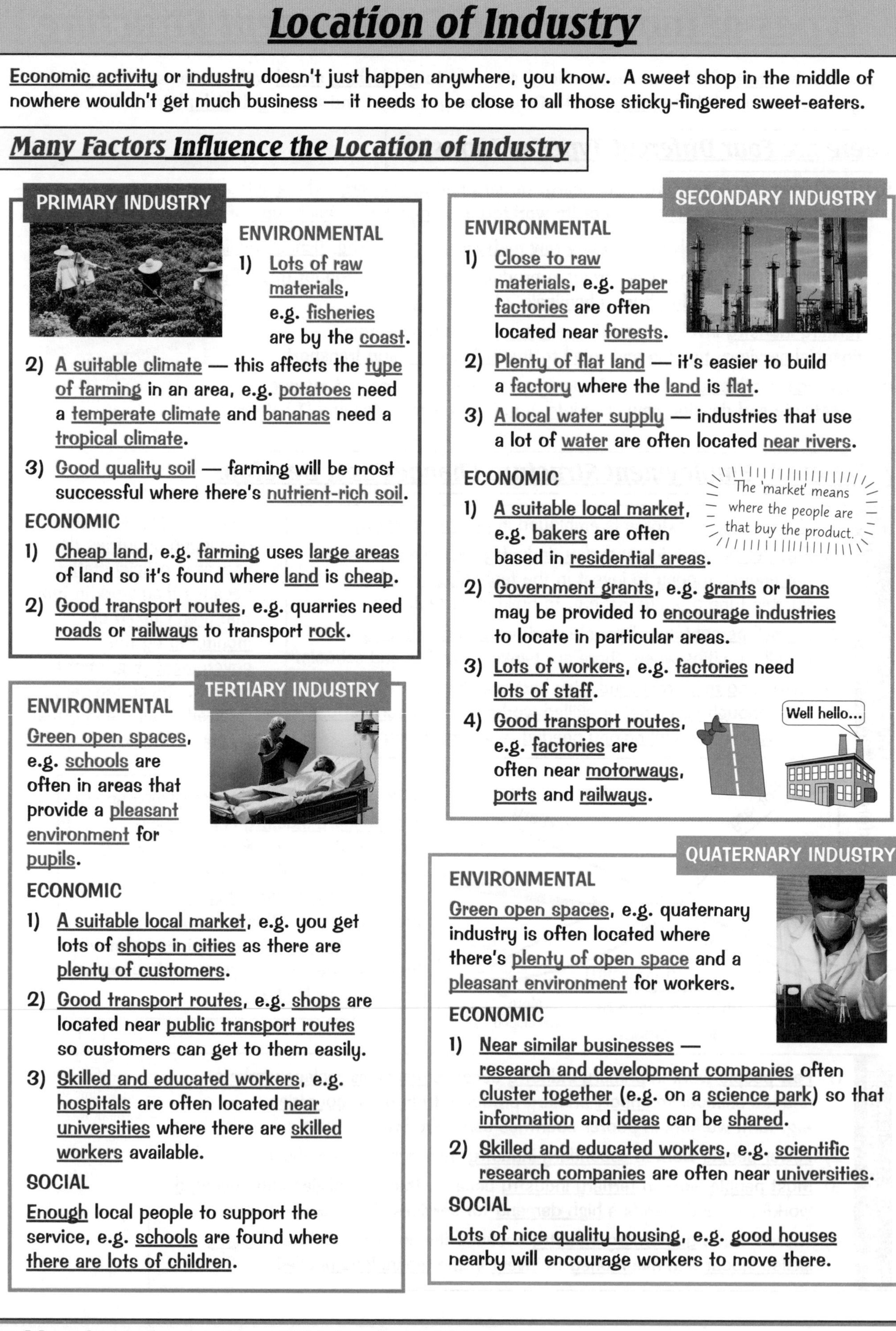

Location of Industry

Economic activity or industry doesn't just happen anywhere, you know. A sweet shop in the middle of nowhere wouldn't get much business — it needs to be close to all those sticky-fingered sweet-eaters.

Many Factors Influence the Location of Industry

PRIMARY INDUSTRY

ENVIRONMENTAL

1) Lots of raw materials, e.g. fisheries are by the coast.
2) A suitable climate — this affects the type of farming in an area, e.g. potatoes need a temperate climate and bananas need a tropical climate.
3) Good quality soil — farming will be most successful where there's nutrient-rich soil.

ECONOMIC

1) Cheap land, e.g. farming uses large areas of land so it's found where land is cheap.
2) Good transport routes, e.g. quarries need roads or railways to transport rock.

SECONDARY INDUSTRY

ENVIRONMENTAL

1) Close to raw materials, e.g. paper factories are often located near forests.
2) Plenty of flat land — it's easier to build a factory where the land is flat.
3) A local water supply — industries that use a lot of water are often located near rivers.

ECONOMIC

1) A suitable local market, e.g. bakers are often based in residential areas.

The 'market' means where the people are that buy the product.

2) Government grants, e.g. grants or loans may be provided to encourage industries to locate in particular areas.
3) Lots of workers, e.g. factories need lots of staff.
4) Good transport routes, e.g. factories are often near motorways, ports and railways.

Well hello...

TERTIARY INDUSTRY

ENVIRONMENTAL

Green open spaces, e.g. schools are often in areas that provide a pleasant environment for pupils.

ECONOMIC

1) A suitable local market, e.g. you get lots of shops in cities as there are plenty of customers.
2) Good transport routes, e.g. shops are located near public transport routes so customers can get to them easily.
3) Skilled and educated workers, e.g. hospitals are often located near universities where there are skilled workers available.

SOCIAL

Enough local people to support the service, e.g. schools are found where there are lots of children.

QUATERNARY INDUSTRY

ENVIRONMENTAL

Green open spaces, e.g. quaternary industry is often located where there's plenty of open space and a pleasant environment for workers.

ECONOMIC

1) Near similar businesses — research and development companies often cluster together (e.g. on a science park) so that information and ideas can be shared.
2) Skilled and educated workers, e.g. scientific research companies are often near universities.

SOCIAL

Lots of nice quality housing, e.g. good houses nearby will encourage workers to move there.

It'd be nice to have a biscuit factory round here...

Bit of a complicated page, this — industry is all about location, location, location. And because it's complicated you need to make extra sure you know it. Use the old drill — cover this page and scribble down as much as you can.

Location of Industry — Case Studies

If you're not feeling very industrious, I think I can help you. Two case studies on why industry locates where it does should boost your enthusiasm — then you can turn this raw material into some exam fodder.

The Location of Industry in a Poorer Country — Kenya

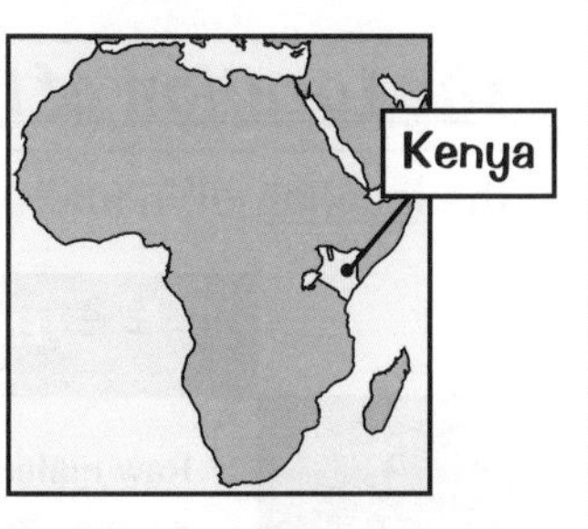

Secondary — there are lots of manufacturers in Nairobi that produce textiles, foods and drinks. The area has good transport links (including an airport) and a good labour supply — Nairobi's population is about 3 million people.

Primary — there are lots of farms in the Nyanza and Western Provinces that produce coffee, tea, tobacco and fruits — these are the areas that receive enough rainfall to grow crops.

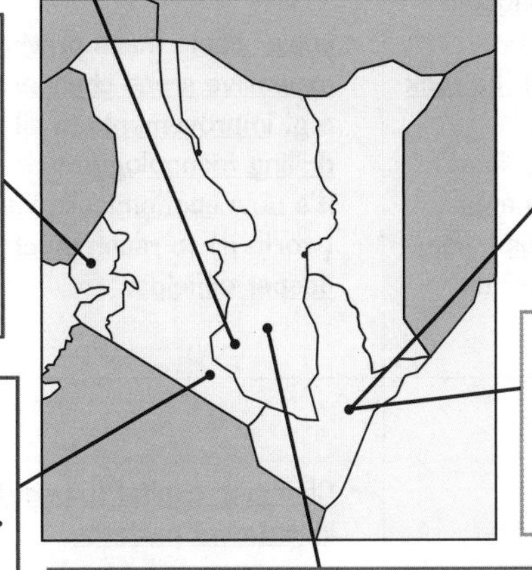

Tertiary — there's a strong tourist industry in the Coast Province because of its beaches, e.g. Diani Beach.

Secondary — there are cement works in the Coast Province because they use limestone from the nearby deposits as a raw material.

Primary — there's a large mine in Magadi that extracts trona (a mineral that's used to make glass). The mineral forms around Lake Magadi as the water evaporates.

Tertiary — there's a strong tourist industry near Mount Kenya because it's the second highest mountain in Africa and has a National Park.

The Location of Industry in a Richer Country — the UK

Quaternary — many electronics companies are based in the Central Lowlands of Scotland because of the local skilled labour supply — nearby universities such as Glasgow, Edinburgh and Heriot-Watt provide electronics and engineering graduates.

Secondary — there are chemicals works in North East England because they're near to offshore oil rigs that provide the raw material for the industry.

Primary — there are lots of farms in Lincolnshire and East Anglia because of the good soil and mild climate. It's also very flat, which makes it easier to use large machinery such as tractors.

Tertiary — there's a strong tourist industry in Cumbria because of the beautiful scenery.

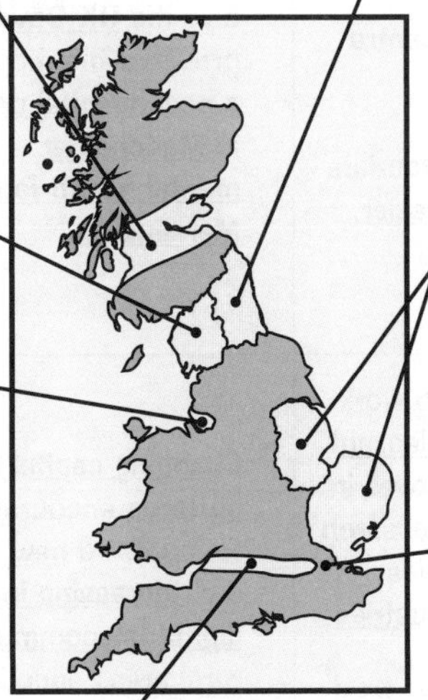

Secondary — government grants have encouraged car manufacturers to locate in Merseyside, e.g. in 1998 Jaguar cars began production at the Halewood plant after receiving a £50 million grant from the Government.

Tertiary — there are lots of shops in London because there are plenty of customers and good public transport. There are also excellent transport links for the delivery of products to shops.

Quaternary — there are many high technology industries along the M4 corridor (the area near the M4 motorway). The area is close to universities (Bristol, Oxford and Reading) that provide skilled workers, and the motorway is good for communication.

Case studies and London buses — you wait for ages and two come along at once...

So if you like a touch of farming, the east of England's the place to be. Industry within your local town will be based in suitable places for similar reasons. Try listing some local companies and working out why they might be based there.

Location of Industry Over Time

Time changes everything, honest. It even changes where industry locates — and not just from the revision chair to the exam room. To help you keep up with all the changes, I've made a table.

The Location of Industry Changes Over Time

The location of industry changes with time due to environmental, economic and social reasons:

	ENVIRONMENTAL	ECONOMIC	SOCIAL
Primary Industry	• Raw materials become exhausted so industry moves elsewhere, e.g. quarries move once all the rock has been extracted. • Climate change means that some crops can be grown in new areas, e.g. increasing temperatures mean that vineyards have been set up in Kent.	• Lower costs make previously expensive areas cheaper, e.g. improvements in oil drilling technology mean that it's now economically viable (worth it) to reach much deeper oilfields.	• Improved transport routes mean that primary industry can be located in more remote areas, e.g. better roads in Brazil mean forestry is possible in new parts of the Amazon rainforest. • Government policies change to allow industry in new areas, e.g. in 2008, Western Australia lifted its ban on uranium mining.
Secondary Industry	• New energy sources mean that industry doesn't have to be close to power sources, e.g. in the past many factories used coal for power so were near coalfields, but now they use electricity from the National Grid.	• Changing capital (money) investment patterns encourage industry to locate to new areas, e.g. the UK Government and private investment in manufacturing in Scotland is encouraging industries to locate there.	• Government policies change, which encourages industries to settle in different locations, e.g. the UK Government gives incentives to companies to open factories (and create jobs) in deprived areas. • Improved transport facilities mean more people have access to cars or public transport so can travel further to work, e.g. in the past many factories were located in city centres so workers could get there easily.
Tertiary Industry	• Workers increasingly want a nice working environment with pleasant surroundings, so industry moves in order to attract workers, e.g. offices move from the centre of a city to the outskirts. • Extreme environments are becoming more popular for tourists as travel gets cheaper and easier, e.g. the tourist industry is developing in Antarctica.	• Changing capital investment patterns encourage industry to locate to new areas, e.g. the UK Government has provided the money to build a new hospital development in Manchester and new mental health facilities in Merseyside.	• Improved transport facilities mean retailers don't have to be located in city centres for their customers to reach them, e.g. most people have access to cars now so there are more out-of-town shopping centres. • Shopping patterns have changed so people don't just shop on their local high street, e.g. many retailers sell products over the internet, so they don't need to be near their customers.
Quaternary Industry	• Workers increasingly want to work in a nice environment with pleasant surroundings, so industry moves in order to attract them, e.g. research centres are often outside cities. • Some scientific research industries have environmental needs, e.g. research into GM crops needs land to grow experimental crops away from ordinary crops.	• Changing capital investment patterns encourage industry to locate to new areas, e.g. increasing investment in digital telecommunications in rural areas encourages businesses to move there.	• The labour force moves as training and housing changes, e.g. electronics industries often locate near universities that have good electronics courses.

Any minute now the joke factory will move to my office...

...and then I'll be able to write something hilarious. While you're waiting, cover up this page and see how much of this table you can scribble down. Then, if you check back in a few minutes I may have something witty for you as a reward.

Environmental Impacts of Industry

Ah, impacts, impacts. As you might have guessed by now, geographers are obsessed by them.

Primary Industries Have a Huge Impact on the Environment

FARMING

Biodiversity is the number and variety of organisms. A habitat is where an organism lives.

1) Monoculture (growing just one type of crop) reduces biodiversity as there are fewer habitats.
2) Removing hedgerows to increase the area of farmland destroys habitats. It also increases soil erosion (hedgerows normally act as windbreaks).
3) Herbicides can kill wildflowers, pesticides can kill other insects (as well as pests) and fertilisers can pollute rivers, killing fish (this is called eutrophication).
4) Making fertilisers, pesticides and herbicides uses fossil fuels, which adds to global warming.
5) Cows produce methane, which also adds to global warming.

There's loads more about global warming on pages 27-29.

MINING

1) Mining destroys large areas of land, so there are fewer habitats and food sources for animals and birds. This reduces biodiversity.
2) Mining uses lots of water, so it can deplete water sources.
3) Some kinds of mining can cause water pollution.

FISHING

1) Overfishing depletes resources and upsets food chains.
2) Fishing boats can leak oil and diesel, which kills aquatic animals.

FORESTRY

1) Fewer trees means fewer habitats and food sources for animals and birds. This reduces biodiversity.
2) Soil erosion is more common as there are fewer trees to hold the soil together.
3) Trees remove CO_2 from the atmosphere when they photosynthesise, so without them less CO_2 is removed. More forestry means more CO_2 in the atmosphere, which adds to global warming.
4) Without trees, less water is removed from the soil and evaporated into the atmosphere. So fewer clouds form and rainfall in the area is reduced. Reduced rainfall reduces plant growth.

Secondary Industries Cause Pollution

1) Factories can cause land, air and water pollution, e.g. dyes from textile factories can pollute rivers and sulfur dioxide emissions from metal works can cause acid rain.
2) Habitats are destroyed if factories are built in the countryside.
3) Some factories use a huge amount of energy, e.g. ice cream factories. This energy usually comes from burning fossil fuels, so adds to global warming.

Tertiary and Quaternary Industries Use a Lot of Energy

1) Tertiary and quaternary industries use a lot of energy, e.g. to run computers, shops or vehicles.
2) This energy usually comes from burning fossil fuels, so adds to global warming.
3) Also, all the resources these industries use cause an impact when they're manufactured. E.g. trees are cut down and made into paper in factories.

Maybe I'll turn my computer off standby then...

Crikey, industry has a lot of environmental impacts (even more impacts than I have skateboarding bruises). Check that you know them so you can trot them out in neat handwriting on demand (if it's illegible you won't get any marks).

Development and Environmental Impacts

Industry is good for a country's pocket (it helps its economy to develop), but it's not great for its back garden. Fear not though my friend, there are things that can be done to make the garden a beautiful oasis.

Economic Development often Damages the Environment

1) An increase in industry in an area helps it to develop economically — it creates more jobs, which increases the wealth of the area and the local people.
2) But some industries damage the environment a lot.
3) This means there's conflict between economic development (by increasing industry) and protecting the environment.
4) Economic development can aim to be sustainable though. To be sustainable it has to increase the wealth of an area in a way that doesn't stop people in the future getting what they need. Basically this means not depleting resources or damaging the environment irreversibly.

There are Ways to Make Economic Development More Sustainable

Economic development can be more sustainable if the industries that cause it reduce their environmental impacts. For example:

FARMING

1) Use fewer herbicides, pesticides and fertilisers (although this reduces crop yield).
2) Maintain hedgerows instead of removing them.

MINING

1) Laws can be introduced to help reduce water pollution.
2) The habitats in a quarry can be restored once it's disused, e.g. by planting trees and creating ponds.

FORESTRY

Laws can be introduced that make logging companies plant one tree for each one cut down. This means that there will still be trees for the future. It also reduces soil erosion.

FISHING

1) Quotas (limits on the number of fish caught) can be introduced to stop overfishing.
2) Fish can be raised on fish farms to prevent wild stocks from running out.

FACTORIES

1) Laws to reduce water, air and land pollution can be introduced.
2) Building on brownfield sites (derelict areas that have been used, but aren't being used any more) stops habitat destruction.
3) Energy use can be reduced by using more energy efficient devices.

OFFICES, SHOPS and VEHICLES

1) Turning off computers instead of leaving them on standby reduces energy use.
2) Using more efficient vehicles reduces the amount of fossil fuel burnt.

I have an ongoing conflict with my brother about chips...

He thinks ketchup is the sauce of choice for chips. But I (correctly) believe mayonnaise is far superior. However, on fish fingers (made from sustainably farmed stocks), there's no such conflict. It's ketchup all the way. Obviously.

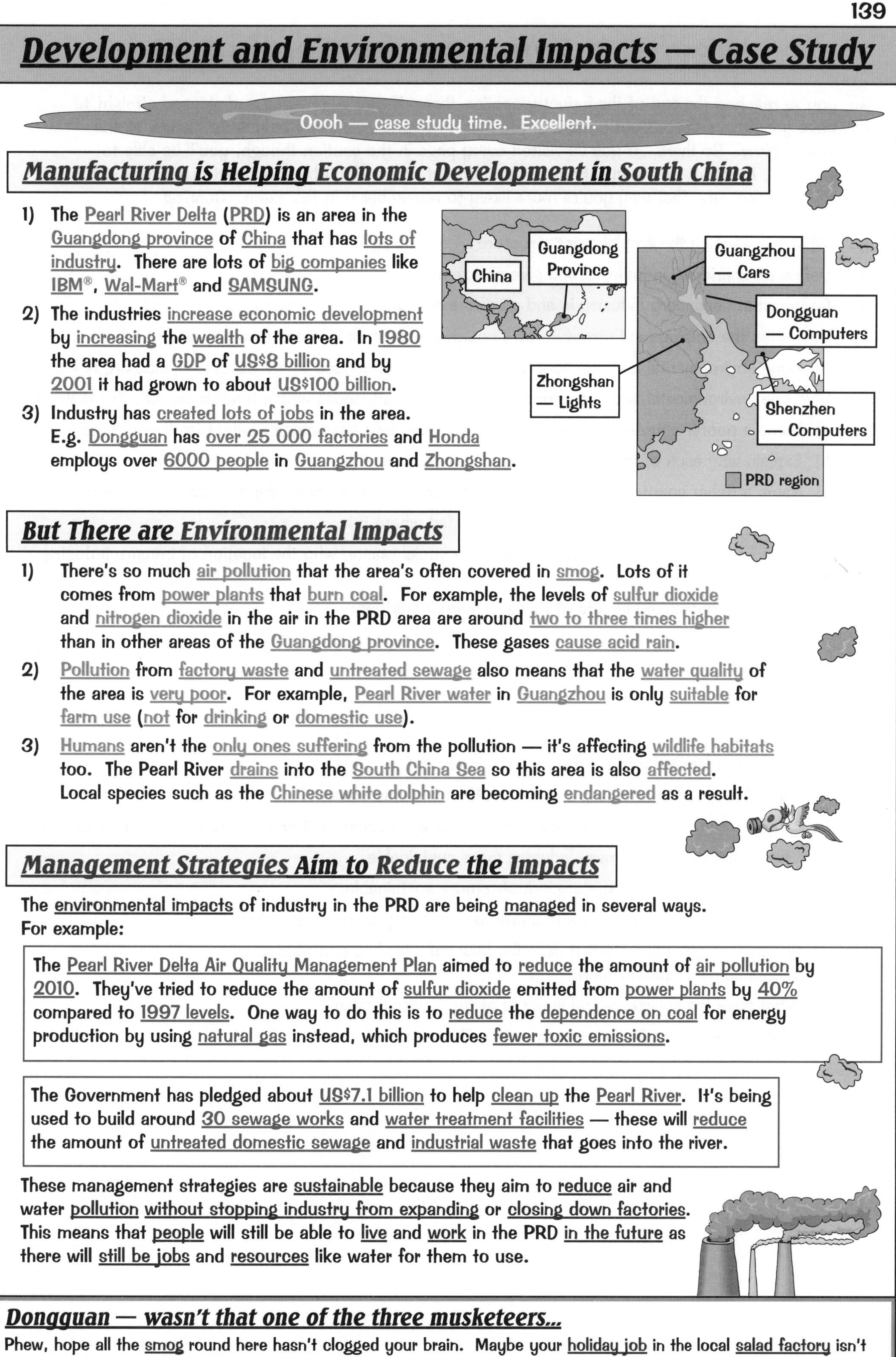

Development and Environmental Impacts — Case Study

Oooh — case study time. Excellent.

Manufacturing is Helping Economic Development in South China

1) The Pearl River Delta (PRD) is an area in the Guangdong province of China that has lots of industry. There are lots of big companies like IBM®, Wal-Mart® and SAMSUNG.
2) The industries increase economic development by increasing the wealth of the area. In 1980 the area had a GDP of US$8 billion and by 2001 it had grown to about US$100 billion.
3) Industry has created lots of jobs in the area. E.g. Dongguan has over 25 000 factories and Honda employs over 6000 people in Guangzhou and Zhongshan.

But There are Environmental Impacts

1) There's so much air pollution that the area's often covered in smog. Lots of it comes from power plants that burn coal. For example, the levels of sulfur dioxide and nitrogen dioxide in the air in the PRD area are around two to three times higher than in other areas of the Guangdong province. These gases cause acid rain.
2) Pollution from factory waste and untreated sewage also means that the water quality of the area is very poor. For example, Pearl River water in Guangzhou is only suitable for farm use (not for drinking or domestic use).
3) Humans aren't the only ones suffering from the pollution — it's affecting wildlife habitats too. The Pearl River drains into the South China Sea so this area is also affected. Local species such as the Chinese white dolphin are becoming endangered as a result.

Management Strategies Aim to Reduce the Impacts

The environmental impacts of industry in the PRD are being managed in several ways. For example:

The Pearl River Delta Air Quality Management Plan aimed to reduce the amount of air pollution by 2010. They've tried to reduce the amount of sulfur dioxide emitted from power plants by 40% compared to 1997 levels. One way to do this is to reduce the dependence on coal for energy production by using natural gas instead, which produces fewer toxic emissions.

The Government has pledged about US$7.1 billion to help clean up the Pearl River. It's being used to build around 30 sewage works and water treatment facilities — these will reduce the amount of untreated domestic sewage and industrial waste that goes into the river.

These management strategies are sustainable because they aim to reduce air and water pollution without stopping industry from expanding or closing down factories. This means that people will still be able to live and work in the PRD in the future as there will still be jobs and resources like water for them to use.

Dongguan — wasn't that one of the three musketeers...

Phew, hope all the smog round here hasn't clogged your brain. Maybe your holiday job in the local salad factory isn't so bad after all — at least there's some nice coleslaw-scented air to breathe. Although that might not be your thing.

Revision Summary for Section 12

So, you've reached the end of the industry section then. If you've cheated and skipped straight to this page you'll have missed out on lots of my bad jokes and you won't be able to answer all the questions below. So there. If you've visited every page in the section though, you'll be able to answer most, if not all, of the questions. If you do get stuck, don't be shy about going back and checking the answer. That way you're more likely to remember it in the exam. Cunning.

1) What is meant by the employment structure of a country?
2) Define what primary industry is and give an example.
3) Define what secondary industry is and give an example.
4) How does the employment structure of a country change as it becomes more developed?
5) Give one environmental and one economic factor that influences where secondary industry is located.
6) Give one environmental and one economic factor that influences where quaternary industry is located.
7) a) Name a poorer country you have studied and give four examples of industries found there.
   b) Explain why each of those industries is located where it is.
8) a) Name a richer country you have studied and give four examples of industries found there.
   b) Explain why each of those industries is located where it is.
9) Give one environmental, one economic and one social reason why the location of primary industry changes over time.
10) Give two environmental impacts of these primary industries:
    a) Farming
    b) Mining
    c) Fishing
    d) Forestry
11) How do secondary industries affect the environment?
12) How do tertiary and quaternary industries affect the environment?
13) Explain why there's often conflict between economic development and protecting the environment.
14) Give two ways to make primary industry more sustainable.
15) Give two ways to make secondary industry more sustainable.
16) Give two ways to make tertiary and quaternary industry more sustainable.
17) a) For an area you have studied describe how industry has contributed to economic development there.
    b) What have the environmental impacts of the industry been?
    c) What is being done in the area to reduce the impacts?

Globalisation Basics

Globalisation is a long word and a complicated subject. Better get started then...

Globalisation is the Process of Economies Becoming More Integrated

1) Globalisation is the process of all the world's economies becoming integrated — it's the whole world coming together like a single community.
2) It happens because of international trade, international investment and improvements in communications.
3) Countries have become interdependent as a result of globalisation — they rely on each other for resources or services.

Improvements in Communications have Increased Globalisation

Improvements in ICT (Information and Communication Technology) and transport have increased globalisation by increasing trade and investment:

ICT

1) Improvements in ICT include e-mail, the internet, mobile phones and phone lines that can carry more information and faster.
2) This has made it quicker and easier for businesses all over the world to communicate with each other. For example, a company can have its headquarters in one country and easily communicate with branches in other countries. No time is lost so it's really efficient.

Transport

1) Improvements in transport include more airports, high-speed trains and larger ships.
2) This has made it quicker and easier for people all over the world to communicate with each other face to face.
3) It's also made it easier for companies to get supplies from all over the world, and to distribute their product all over the world. They don't have to be located near to their suppliers or their product market anymore.

These improvements have allowed the development of call centres abroad and localised industrial regions:

Call centres abroad

1) Call centres are used by some companies to handle telephone enquiries about their business.
2) Improvements in ICT mean that it's just as easy for people to phone a faraway country as it is to phone people in their own country.
3) So a lot of call centres are now based abroad because the labour is cheaper, which reduces running costs.

EXAMPLE: In 2004 Aviva (an insurance company) moved 950 call centre jobs from the UK to India and Sri Lanka, as it costs less there (e.g. it costs 40% less in India).

Localised industrial regions

Improvements in ICT and transport have allowed some industries to develop around a specific region that's useful to them, but still have global connections to get all the other things they need.

EXAMPLE: A lot of motorsport companies have offices in Oxfordshire and Northamptonshire, e.g. the Renault Formula 1 team have their headquarters there. They're close to the Silverstone race circuit (so they can test their cars) and the area has lots of skilled workers. People like drivers and engineers can easily fly into the area. The manufacturers use the internet to easily send and receive information and data about their cars to people around the world.

Globalisation is going large — with extra barbecue sauce...

Globalisation is a bit of a weird concept so don't panic if you don't get it straight away. This page has a lot of information on it so just take it slowly — make sure you understand it before moving on to the rest of the section.

Trans-National Corporations (TNCs)

You can't get too far into the topic of globalisation before stumbling across Trans-National Corporations.

TNCs also Increase Globalisation

1) TNCs are companies that produce products, sell products or are located in more than one country. For example, Sony is a TNC — it manufactures electronic products in China and Japan, and sells many of them in Europe and the USA.
2) TNCs are usually very rich companies that employ lots of people and have a large output (they make loads of products every year). For example:

Ford is an American-owned TNC that makes cars. In 2008, it produced over 5 million cars worldwide. It also employs over 200 000 people at about 90 different sites around the world.

3) TNCs increase globalisation by linking together countries through the production and sale of goods.
4) They also bring the culture from their country of origin to many different countries, e.g. McDonald's brings Western-style fast food to other countries.

TNCs are also known as multinational companies (MNCs).

TNCs Affect Economic Development

1) TNCs create jobs in an area. This increases the wealth of the area (due to taxes) and the wealth of the local people (due to employment).
2) Taxes are used to improve infrastructure (e.g. roads) and services (e.g. schools, hospitals, etc.). People also have more money to spend. Both of these things attract more businesses to the area (including more TNCs), creating even more jobs, and so on...
3) This cycle (more jobs, leading to more services, leading to more jobs...) is called the multiplier effect.
4) TNC factories are often located in poorer countries because labour is cheaper, which means they make more profit. (See page 144 for more reasons why they're located in poorer countries.)
5) TNC headquarters and research centres are usually located in richer countries because there are more skilled and educated people (but there are some TNC factories in richer countries as well).

TNCs have Advantages and Disadvantages

Advantages	Disadvantages
TNCs create jobs in all the countries they're located in.	Employees in poorer countries may be paid lower wages than employees in richer countries.
Employees in poorer countries get a more reliable income compared to jobs like farming.	Employees in poorer countries may have to work long hours in poor conditions.
When they locate to poorer countries, TNCs create some skilled jobs, e.g. jobs in factory offices. This encourages more education and training in the area.	Most TNCs come from richer countries so the profits go back there — they aren't reinvested in the poorer countries they operate in.
TNCs spend money to improve the local infrastructure, e.g. airports and roads.	Large sites will attract lots of traffic, which increases pollution in the area.
New technology (e.g. computers) and skills are brought to poorer countries.	The jobs created in poorer countries aren't secure — the TNC could relocate the jobs to another country at any time.
Local companies supply the TNCs, increasing their income.	Other local companies may struggle to find business or workers, so shut down.

TNCs are everywhere — and I mean everywhere...

Geography types are always making up long names for things (trans-national corporations), then squishing them down to just a few letters (TNC). I don't know what they get out of it but it seems to keep them happy.

TNCs — Case Study

FACT: only cockroaches and case studies can survive a direct nuclear explosion.

Wal-Mart® is a Retail TNC with Headquarters in the USA

1) Wal-Mart began in 1962 when Sam Walton opened the first store in Arkansas, USA.
2) More stores opened across Arkansas, then across the USA, and more recently across the world, e.g. in Mexico, Argentina, China, Japan, Brazil, Canada and the UK (where it's called ASDA).
3) Wal-Mart sells a variety of products, e.g. food, clothes and electrical goods.
4) Wal-Mart is the biggest retailer in the world — it owns over 8000 stores and employs over 2 million people.

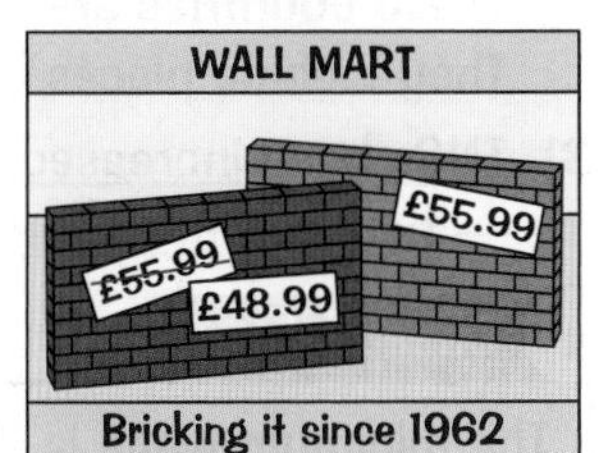

Bricking it since 1962

Wal-Mart has Positive Effects...

■ = Location of Wal-Mart stores

1) Wal-Mart creates lots of jobs in different countries, e.g. in construction, manufacturing and retail services. E.g. in Mexico, Wal-Mart employs over 150 000 people and in Argentina, three new stores opened in 2008, creating nearly 450 jobs.

2) Local companies and farmers supply goods to Wal-Mart, increasing their business. E.g. in Canada, Wal-Mart works with over 6000 Canadian suppliers, creating around US$11 billion of business for them each year.

3) Wal-Mart offers more skilled jobs in poorer countries. E.g. all the Wal-Mart stores in China, are managed by local people.

4) Wal-Mart donates hundreds of millions of dollars to improve things like health and the environment in countries where it's based. E.g. in 2008 in Argentina, Wal-Mart donated US$77 000 to local projects and gave food and money to help feed nearly 12 000 poor people.

5) The company invests money in sustainable development. E.g. in Puerto Rico, 23 Wal-Mart stores are having solar panels fitted on their roofs to generate electricity.

...and Negative Effects

1) Some companies that supply Wal-Mart have long working hours. E.g. Beximco in Bangladesh supplies clothing. Bangladesh has a maximum 60 hour working week, but some people claim employees at Beximco regularly work 80 hours a week.
2) Not all Wal-Mart workers are paid the same wages. E.g. factory workers in the USA earn around $6 an hour, but factory workers in China earn less than $1 an hour (although this is quite a lot in China).
3) Some studies have suggested that Wal-Mart stores can cause smaller shops in the area to shut — they can't compete with the low prices and range of products on sale.
4) The stores are often very large and out-of-town, which can cause environmental problems. Building them takes up large areas of land and people driving to them causes traffic and pollution. For example, the largest Wal-Mart store is in Hawaii and it covers over 27 000 m^2 — that's over three times the size of the football pitch at Wembley Stadium.

Wal-Mart — for all your walling needs...

If you know squillions of details for another TNC case study then that's fine (as long as you do know them — owning some of their products doesn't count). If you don't then get your memorising hat on (mine's a pink fez) and get learnin'.

Change in Manufacturing Location

TNCs can put their factories anywhere in the world, but some countries are more attractive than others. This has meant some countries are now stuffed full of factories and others are waving bye bye to them.

The Manufacturing Industry is Growing in Some Countries...

1) Some countries that have traditionally relied on agriculture have seen a massive growth in their manufacturing industries recently — this process is called industrialisation.
2) These countries are called NICs (Newly Industrialising Countries). They include places like India, China and Brazil.
3) TNCs have increased manufacturing in NICs by basing factories there — here are five reasons why they do this:

An increase in manufacturing creates jobs.

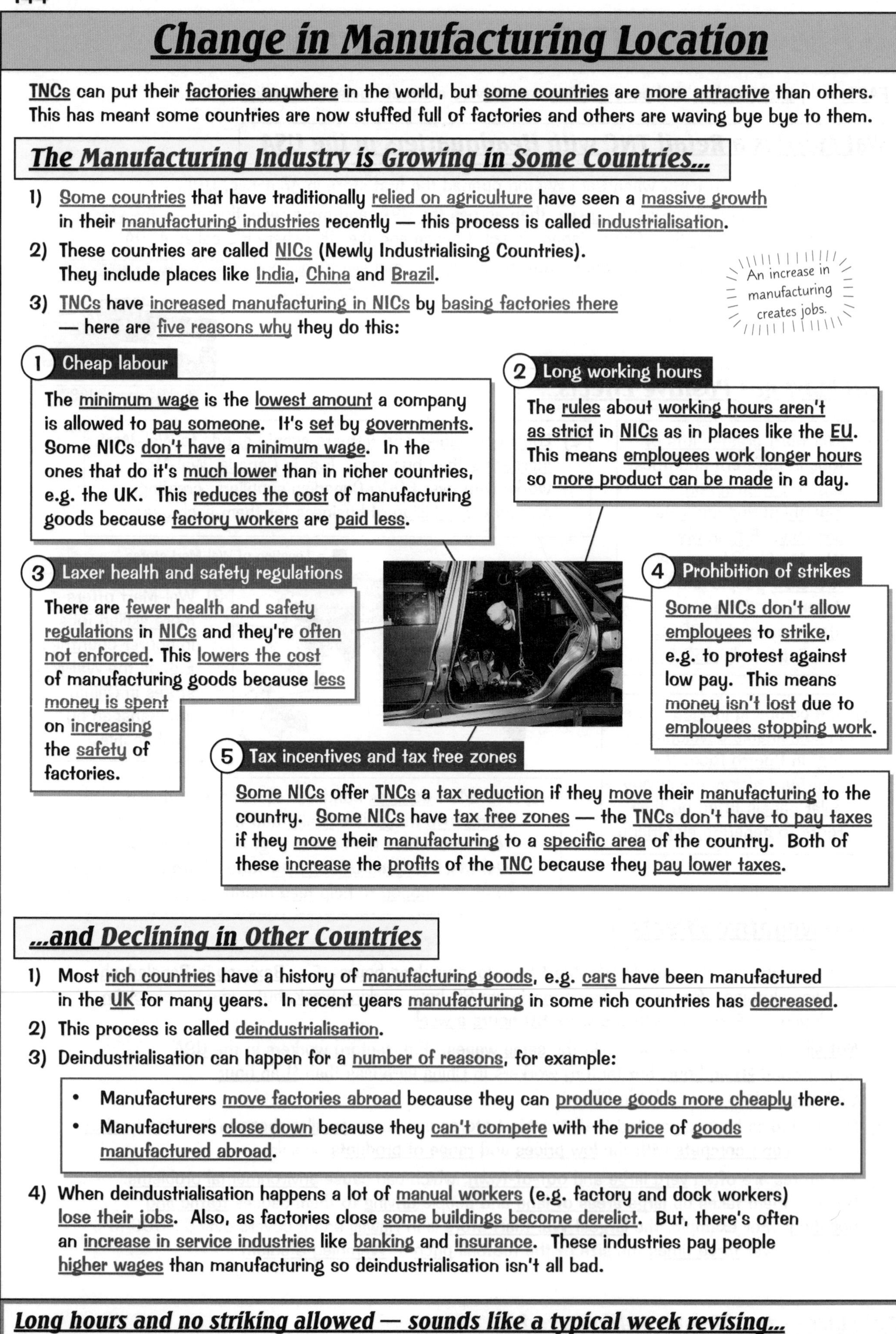

1 Cheap labour

The minimum wage is the lowest amount a company is allowed to pay someone. It's set by governments. Some NICs don't have a minimum wage. In the ones that do it's much lower than in richer countries, e.g. the UK. This reduces the cost of manufacturing goods because factory workers are paid less.

2 Long working hours

The rules about working hours aren't as strict in NICs as in places like the EU. This means employees work longer hours so more product can be made in a day.

3 Laxer health and safety regulations

There are fewer health and safety regulations in NICs and they're often not enforced. This lowers the cost of manufacturing goods because less money is spent on increasing the safety of factories.

4 Prohibition of strikes

Some NICs don't allow employees to strike, e.g. to protest against low pay. This means money isn't lost due to employees stopping work.

5 Tax incentives and tax free zones

Some NICs offer TNCs a tax reduction if they move their manufacturing to the country. Some NICs have tax free zones — the TNCs don't have to pay taxes if they move their manufacturing to a specific area of the country. Both of these increase the profits of the TNC because they pay lower taxes.

...and Declining in Other Countries

1) Most rich countries have a history of manufacturing goods, e.g. cars have been manufactured in the UK for many years. In recent years manufacturing in some rich countries has decreased.
2) This process is called deindustrialisation.
3) Deindustrialisation can happen for a number of reasons, for example:
   - Manufacturers move factories abroad because they can produce goods more cheaply there.
   - Manufacturers close down because they can't compete with the price of goods manufactured abroad.
4) When deindustrialisation happens a lot of manual workers (e.g. factory and dock workers) lose their jobs. Also, as factories close some buildings become derelict. But, there's often an increase in service industries like banking and insurance. These industries pay people higher wages than manufacturing so deindustrialisation isn't all bad.

Long hours and no striking allowed — sounds like a typical week revising...

The basic formula is 'more factories here = less factories there'. Get that fact fixed in your grey matter, then re-read the reasons why it's the height of fashion to put your factory in a NIC. Don't forget about deindustrialisation either.

Change in Manufacturing Location — Case Study

You're not getting away that easily — here's another case study for you. This page is about China's development into an economic giant and the reasons why manufacturing is moving to the country.

China is one of the World's Fastest Growing Industrial Economies

1) In 30 years China has gone from being a mainly agricultural economy to a strong manufacturing economy. It's now the third largest economy in the world after the US and Japan.
2) The percentage of China's GDP that came from agriculture fell between 1978 and 2004, from about 30% to less than 15%.
3) During the same time the number of products manufactured in China has increased rapidly, e.g. about 4000 colour TVs were made in China in 1978 compared to nearly 75 million in 2004.
4) China manufactures loads of different products like clothes, computers and toys.
5) Lots of TNCs have factories in China, for example NIKE, Hewlett-Packard and Disney.

GDP is the total value of goods and services a country produces in a year.

There are Lots of Reasons for the Growth in Manufacturing

1 Cheap labour

There's no single minimum wage in China — it's different all over the country. For example, in Shenzhen the minimum wage is about £90 per month and in Beijing it's about £70 per month. This makes labour in China much cheaper than other countries, e.g. in the UK the minimum wage is about £990 per month.

2 Long working hours

Chinese law says that people are only allowed to work 40 hours per week, with a maximum of 36 hours of overtime per month. This isn't always enforced though — for example, the manufacturing company foxconn® said that some of its Chinese factory workers have done about 80 hours of overtime per month to maximise the production of goods.

3 Laxer health and safety regulations

The health and safety laws in China are similar to other countries but they aren't heavily enforced, e.g. over the past decade, hundreds of factory workers have been treated for mercury poisoning despite strict laws on working with toxic materials.

4 Prohibition of strikes

Chinese workers can go on strike but the All-China Federation of Trade Unions (ACFTU) is required by law to get people back to work as quickly as possible so productivity is maximised. It's illegal for people to join any union other than the ACFTU.

5 Tax incentives and tax free zones

China has many Special Economic Zones (SEZs) that offer tax incentives to foreign businesses. Foreign manufacturers usually pay no tax for the first two years in the zone, 7.5% for the next three years and then 15% from then on (which is still half of the usual 30% tax elsewhere in China). Shenzhen is one of the most successful SEZs. There's been around $30 billion of investment by TNCs. Factories in Shenzhen make products for companies like Wal-Mart®, Dell™ and IBM®.

Fe fi fo fum — China gets the world's manufacturing done...

I bet you're within about a metre of something that was made in China — unless you're doing your revision on the surface of the Sun. Check that you can give all the reasons why manufacturing in China has gone through the roof.

Globalisation and Energy Demand

All this globalisation has made us hungry. No, not for courgettes, but for energy...

The Global Demand for Energy is Increasing

Globalisation has increased the wealth of some poorer countries so people are buying more things. A lot of these things use energy, e.g. cars, fridges and televisions. This increases the global demand for energy.

There are two other reasons why the global demand for energy is increasing:

1) Technological advances have created loads of new devices that all need energy, e.g. computers, mobile phones and MP3 players. These are becoming more popular so more energy is needed.
2) In 2000 the world population was just over 6 billion and it's projected to increase to just over 9 billion in 2050 — more people means more energy is needed.

Producing More Energy has Lots of Impacts

Most of the energy produced in the world comes from burning fossil fuels (i.e. oil, gas and coal). Nuclear power, wood and renewable sources (e.g. solar power) are also used to produce some energy. Increasing energy production to meet demand has social, economic and environmental impacts:

Social impacts

1) More power plants will have to be built to increase energy production. Power plants are extremely large — people may have to move out of an area so a power plant can be built.
2) The waste from nuclear power plants is radioactive. If it leaks out from where it's stored it can cause death and illness, and can contaminate large areas of land. If more nuclear power plants are built to increase energy production, there's a higher risk of radioactive waste leaking out.
3) Increasing energy production will create jobs — people will be needed to build more power stations, run them and maintain them.

Environmental impacts

1) Burning fossil fuels releases carbon dioxide (CO_2). This adds to global warming. Global warming will cause the sea level to rise, cause more severe weather and force species to move (to find better conditions) or make them extinct (if they can't move and it gets too hot). Using more fossil fuels will increase global warming.
2) Burning fossil fuels also releases other gases that dissolve in water in the atmosphere and cause acid rain. Acid rain can kill animals and plants. Using more fossil fuels will increase acid rainfall.
3) Gathering wood for fuel can cause deforestation (removing trees from forests). Removing trees destroys habitats for animals and other plants. Using more wood for fuel will increase deforestation.
4) Mining for coal causes air and water pollution. It also removes large areas of land, which destroys habitats. More coal mining will cause more pollution and destroy more habitats.
5) Transporting oil is a risky business — oil pipes and tankers can leak, spilling oil. Oil spills can kill birds and fish. Using more oil means more needs to be transported, increasing the risk of spills.

Economic impact

Countries with lots of energy resources, e.g. lots of coal, will become richer as energy demand increases — countries with few resources will need to buy energy from them.

The Triffid — a different kind of power plant...

Well, producing energy certainly has a lot of impacts and the impacts are bigger the more energy you produce. Cover the page and see just how many of them you can remember. Go on, give it a go, it might even be fun (but it's rather unlikely).

Globalisation and Food Supply

I'm quite partial to strawberries with my Christmas cake, and it seems I'm not the only one...

Food Production has become Globalised

1) Before the 1960s people mainly ate a small range of seasonal food that had been grown in their own country (often in their local area).
2) People now demand to have a range of foods all year round, regardless of growing seasons. This has led to globalisation of the food industry — food is produced in foreign countries and imported.
3) The increase in the world's population also means more food is needed — the demand has increased.
4) Countries are trying to increase food production to meet this demand, but some can't produce enough to feed their population so food has to be imported too.

Producing More Food and Importing Food has Lots of Impacts

Environmental

1) Transporting food produces CO_2. The distance food is transported to the market is called food miles. The higher the food miles, the more CO_2 is produced. CO_2 adds to global warming.
2) The amount of CO_2 produced during growing and transporting a food is called its carbon footprint. A larger carbon footprint means more CO_2 and more global warming.
3) Imported foods have to be transported a long way so have high food miles and a large carbon footprint. A benefit of importing food is that a wide range of food is available all year round. Another benefit is it helps meet increasing demand in countries that can't produce a lot.
4) More food could be produced locally by energy intensive farming — pesticides, fertilisers and machinery are used to produce large quantities of food. Although food miles are low, loads of energy is needed to make chemicals and run the machinery. Energy production creates lots of CO_2 so local energy intensive farming can have a large carbon footprint.
5) To produce more food some farmers use marginal land (land that's not really suitable for farming), e.g. steep hillsides or the edges of deserts. The soil in marginal land is thin and it's quickly eroded by farming, degrading the environment.

Political

Lots of water is needed to produce lots of food. Farmers in countries with low rainfall need to irrigate their land with water from rivers and lakes. As the demand for water increases (due to the increased demand for food) there may be hostilities between countries that use the same water source for irrigation. For example, there's tension between Egypt, Sudan and Ethiopia because they all take water from the River Nile.

Social

Some farmers are switching from subsistence farming (where food is produced for their family) to commercial farming (where food is produced to sell). This is because they can make more money due to the high demand for food. This reduces the amount of food produced for local people so they have to import food (which is more expensive). If food prices go down, then farmers might not earn enough money to buy food for themselves.

Crops sold to make money are called cash crops.

Economic

1) Using chemicals (e.g. fertilisers, pesticides and insecticides) helps to produce lots of food. These chemicals can be very expensive — farmers may have to borrow money to buy the chemicals and this gets farmers into debt.
2) Farmers can generate a steady income by producing food for export to other countries.

Eat globally — don't stop till you're a perfect sphere...

Farming used to be a totally local business but nowadays our food comes flying in from all over the world (or trundling in on a lorry, train or container ship). There are more mouths to feed too. Check that you know all the knock-on effects.

Reducing the Impacts of Globalisation

One impact of globalisation is that more people are gluttons for energy. Producing more energy using fossil fuels has plenty of impacts, but worry not, things can be done to make producing energy greener.

Using Renewable Energy Is a Sustainable Way to Meet Energy Demands

1) Energy production needs to be sustainable — it needs to allow people alive today to get what they need (energy), but without stopping people in the future getting what they need. This basically means not damaging the environment or using up resources faster than they can be replaced.
2) Producing energy using fossil fuels (i.e. coal, oil and gas) isn't sustainable.
3) This is because fossil fuels are non-renewable — this means they'll eventually run out so there won't be any for future generations.
4) Using fossil fuels also damages the environment, e.g. burning them produces CO_2, which causes global warming.
5) Energy produced from renewable sources is sustainable because it doesn't cause long term environmental damage and the resource won't run out. Here are some renewable energy sources:

Wander back to p. 146 for more on the impacts of producing energy.

- Wind — the wind turns blades on a wind turbine to generate electrical energy.
- Biomass — biomass is material that comes from organisms that are alive (e.g. animal waste) or were recently alive (e.g. plants). It can be burnt to release energy. It can also be processed to produce biofuels, which are then burnt to release energy.
- Solar power — energy from the sun can be used to heat water, cook food and generate electrical energy.
- Hydroelectric power — water is trapped behind a dam and forced through tunnels. The water turns turbines in the tunnels to generate electrical energy.

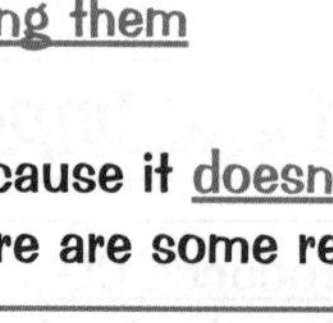

6) Producing energy from renewable sources contributes to sustainable development — it allows areas to develop (i.e. use more energy to improve the lives of the people there) in a sustainable way.

Case Study — Spain is Using Wind Energy to Meet Demand

1) Spain's energy consumption has increased 66% since 1990.
2) Some of the extra energy needed is being produced using wind turbines — the amount of energy produced from wind has increased 16-fold since 1995.
3) Spain is ideal for wind farms because it has large, windy areas where not many people live. This means wind farms can be built without annoying too many people.
4) Spain has over 400 wind farms and a total of over 12 000 turbines.
5) In 2008, 11.5% of Spain's energy was supplied by wind energy.
6) The wind farms have had positive and negative impacts:

Wind farms are groups of wind turbines.

Positive impacts

1) In 2008, using wind energy reduced Spain's CO_2 emissions by over 20 million tonnes.
2) In 2008, using wind energy saved Spain from importing about €1.2 billion of gas and oil.
3) Spain's wind energy industry has created around 40 000 jobs.

Negative impacts

1) Some conservationists say the wind farms are a danger to migrating birds.
2) Some people think wind farms are ugly — turbines can be seen from miles away.
3) Some people think the wind farms are too noisy.

El stinky — a Spanish wind farm...

Sorry, I've sunk to lowly fart jokes after that page. It's not the most thrilling stuff in the world, but unfortunately it's a good idea to know how renewable energy is sustainable and a case study of its use. Happy revising.

Reducing the Impacts of Globalisation

This is the last page of lovely globalisation — then you can head off to far more exciting things like the physics of time travel, the history of shopping, the geometry of the off-side rule... or just more geography.

The Kyoto Protocol aimed to Reduce Carbon Dioxide Emissions

1) Globalisation has increased the demand for energy (see p. 146) — more fossil fuels are being used to meet the demand, producing loads of CO_2 and adding to global warming.
2) The international community is working together to reduce the amount of CO_2 they produce because the problem of global warming affects everyone.
3) The Kyoto Protocol was an international agreement that was signed by most countries in the world to cut emissions of CO_2 and other gases by 2012. Each country was set an emissions target, e.g. the UK agreed to reduce emissions by 12.5% by 2012.
4) Another part of the protocol was the carbon credits trading scheme:

- Countries that came under their emissions target got carbon credits which they could sell to countries that didn't meet their emissions target. This meant there was a reward for having low emissions.
- Countries could also earn carbon credits by helping poorer countries to reduce their emissions. The idea was that poorer countries would be able to reduce their emissions more quickly.

The Kyoto Protocol was due to expire in 2012, but many countries agreed to extend it to 2020.

International agreements are also called international directives.

Other International Agreements help to Reduce Pollution

1) Globalisation has increased the emission of gases that cause pollution like acid rain (see p. 146).
2) There are international agreements that help to reduce pollution, e.g. the Gothenburg Protocol.
3) The Gothenburg Protocol set emissions targets for European countries and the US. The protocol aimed to cut harmful gas emissions by 2010 to reduce acid rain and other pollution.

Recycling Reduces Waste Created by Globalisation

1) Globalisation means people have access to more products at low prices, so they can afford to be more wasteful, e.g. people throw away damaged clothes instead of repairing them.
2) Things that are thrown away get taken to landfill sites — the amount of waste going to landfill has increased as globalisation has increased.
3) One way to reduce this impact on a local scale is to recycle waste to make new products, e.g. recycling old drinks cans to make new ones.

Buying Local can Reduce the Impacts of a Globalised Food Supply

1) In recent years celebrity chefs, food writers and campaigners have encouraged people to eat more locally-produced food.
2) Buying local food helps to reduce food miles (see p. 147) because it hasn't been transported a long way. It also helps to support local farmers and businesses.
3) However, if people only buy locally it can put people in poorer countries who export food out of a job.

You're not local, are you...

And the Oscar® for the best directive goes to — the Gothenburg Protocol...

You know I said that once you've done this page, that's it for globalisation... well, it might have been a tiny lie. When you've learnt this page you can scoot on to the revision summary. After that, I promise you the section is done...

Revision Summary for Section 13

Globalisation... not the most cheery or thrilling of sections (for drama and excitement give me volcanoes any day of the week) — but still kind of important for life, the future of the planet and all that jazz. Before you put it all behind you and move on to something more uplifting, check you've got the hang of the main issues with this useful bunch of bananas. I mean questions. Sorry.

1) What is globalisation?
2) How have improvements in ICT increased globalisation?
3) How have improvements in transport increased globalisation?
4) Why are call centres often based abroad?
5) What is a TNC?
6) How do TNCs increase globalisation?
7) Describe the multiplier effect.
8) Give two advantages and two disadvantages of TNCs.
9) a) Name a TNC.
   b) Name two countries it is located in.
   c) Give one advantage of the TNC.
   d) Give one disadvantage of the TNC.
10) What is industrialisation?
11) What are NICs? Name one NIC.
12) Give two reasons why TNCs move to NICs.
13) Why does deindustrialisation happen?
14) For an NIC you have studied:
    a) Describe how the economy has changed over the last 30 years.
    b) Give two reasons why manufacturing has increased there.
15) Explain how globalisation has increased the demand for energy.
16) Give two other things that have caused an increase in the demand for energy.
17) Give two environmental impacts of producing more energy.
18) Give an economic impact of producing more energy.
19) Give two reasons why food is imported into a country.
20) Describe a political impact of producing more food.
21) Describe a social impact of switching from subsistence farming to commercial farming.
22) What is sustainable energy production?
23) Why are fossil fuels not a sustainable energy source?
24) Give two examples of renewable energy sources.
25) a) Name a country that uses renewable energy.
    b) How much of the energy used in that country comes from renewable sources?
    c) Give one positive impact on that country of using renewable energy.
26) What is the Kyoto Protocol?
27) How does a country get carbon credits?
28) Name an international agreement that aims to reduce pollution.
29) Globalisation has increased the amount of waste going to landfill. How can this impact be reduced?
30) Give one reason why buying locally produced food reduces the impact of a globalised food supply.

Growth in Tourism

Sometimes, geography can be a cruel mistress. Not only do you have to stay inside and do revision all day, but just to rub it in here's a section about other people going on holiday.

There's been a Global Increase in Tourism Over the Last 60 Years

Tourism's a growing industry — people are having more holidays and longer holidays.
Here are a few of the reasons why:

1) People have more disposable income (spare cash) than they used to, so can afford to go on more holidays.
2) Companies give more paid holidays than they used to. This means people have more free time, so go on holiday more.
3) Travel has become cheaper (particularly air travel) so more people can afford to go on holiday.
4) Holiday providers, e.g. tour companies and hotels, now use the internet to sell their products to people directly, which makes them cheaper. Again, this means more people can afford to go away.

Some areas are also becoming more popular than they used to be because:

1) Improvements in transport (e.g. more airports) have made it quicker and easier to get to places — no more week-long boat trips to Australia for a start.
2) Countries in more unusual tourist destinations like the Middle East and Africa have got better at marketing themselves as tourist attractions. This means people are more aware of them.
3) Many countries have invested in infrastructure for tourism (e.g. better hotels) to make them more attractive to visitors.

Cities, Mountains and Coasts are all Popular Tourist Areas

People are attracted to cities by the culture (e.g. museums, art galleries), entertainment (bars, restaurants, theatres) and shopping. Popular destinations include London, New York, Paris and Rome.

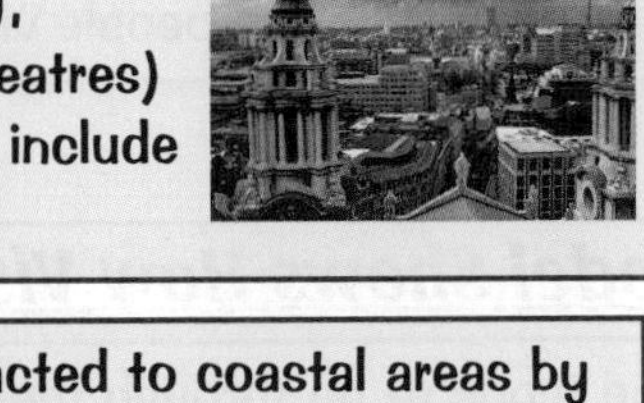

People are attracted to coastal areas by the beaches and activities like swimming, snorkelling, fishing and water skiing. Popular destinations include Spain, the Caribbean and Thailand.

People are attracted to mountain areas by the beautiful scenery and activities like walking, climbing, skiing and snow boarding. Popular destinations include the Alps, the Dolomites and the Rockies.

Tourism is Important to the Economies of Many Countries

1) Tourism creates jobs for local people (e.g. in restaurants and hotels), which helps the economy to grow.
2) It also increases the income of other businesses that supply the tourism industry, e.g. farms that supply food to hotels. This also helps the economy to grow.
3) This means tourism is important to the economy of countries in both rich and poor parts of the world, e.g. tourism in France generated 35 billion Euros in 2006 and created two million jobs.
4) Poorer countries tend to be more dependent on the income from tourism than richer ones, e.g. tourism contributes 3% of the UK's GNP, compared to 15% of Kenya's.

Tourism growth — you should probably have that removed...

Now try to recite the reasons why tourism is on the increase backwards, whilst standing on your head. And don't forget — if an area is pretty or has tons of activities then it'll be a hit with tourists (which is great for the economy).

UK Tourism

You might not realise it on yet another rainy day here, but the UK is actually a top tourist destination.

Tourism makes a Big Contribution to the UK Economy

1) There were 32 million overseas visitors to Britain in 2008.
2) The UK is popular with tourists because of its countryside, historic landmarks (e.g. Big Ben and Stonehenge), famous churches and cathedrals (e.g. Saint Paul's cathedral), and its castles and palaces (e.g. Edinburgh Castle and Buckingham Palace).
3) London is particularly popular for its museums, theatres and shopping. It's the destination for half of all visitors to the UK.
4) In 2007, tourism contributed £114 billion to the economy and employed 1.4 million people.

Really popular areas are called honeypot sites.

Many Factors Affect the Number of Tourists Visiting the UK

1) Weather — bad weather can discourage tourists from visiting the UK, e.g. a really wet summer in 2007 was blamed for a drop in the number of overseas visitors.
2) World economy — in times of recession people tend to cut back on luxuries like holidays, so fewer overseas visitors come to the UK. It's not all bad though, as more UK citizens choose to holiday in the UK.
3) Exchange rate — the value of the pound compared with other currencies affects the number of tourists. If it's low, the UK is cheaper to visit so more overseas visitors come.
4) Terrorism and conflict — wars and terrorist threats mean people are less willing to visit affected areas. Tourism fell sharply after the London bombings on 7th July 2005.
5) Major events — big events can attract huge numbers of people. E.g. Liverpool was European Capital of Culture in 2008 and as a result 3.5 million people visited that hadn't been before.

The Tourist Area Life Cycle Model Shows How Visitor Numbers Change

The number of visitors to an area over time tends to go through these typical stages:

Number of visitors

① Exploration: Small numbers of visitors are attracted to the area, e.g. by the scenery or culture. There aren't many tourist facilities.

② Involvement: Local people start providing facilities for the tourists, which attracts more visitors.

③ Development: More and more visitors come as more facilities are built. Control of tourism in the area passes from locals to big companies.

④ Consolidation: Tourism is still a big part of the local economy, but tourist numbers are beginning to level off.

⑤ Stagnation: Visitor numbers have peaked. Facilities are no longer as good and tourists have had a negative impact on the local environment, making the area less attractive to visit.

⑥ Rejuvenation OR decline:
Rejuvenate — if the area is rejuvenated then more visitors will come as they're attracted by the new facilities.
Decline — fewer visitors come as the area is less attractive. This leads to decline of the area as facilities shut or become run-down.

Time

This model is also known as the resort life cycle model.

I reckon it's the orderly queuing that attracts visitors to the UK...

The tourist area life cycle model applies to many seaside resorts in the UK. They were major tourism centres at the start of the 1900s but many stagnated, e.g. Morecambe, and declined. Some, e.g. Brighton, are now being rejuvenated.

UK Tourism — Case Study

The Lake District National Park has been a favourite with tourists since the early 1800s. Today, the huge number of visitors have to be carefully managed to preserve the natural beauty that's made it so popular.

The Lake District is a National Park in Cumbria

The Lake District National Park gets around 15 million visitors a year.
There are several reasons it's so popular:

1) Tourists come to enjoy the scenery — for example large lakes (e.g. Windermere) and mountains (e.g. Scafell Pike).
2) There are many activities available, e.g. bird watching, walking, pony-trekking, boat rides, sailing and rock-climbing.
3) There are also cultural attractions, e.g. the Beatrix Potter and Wordsworth museums.

Strategies are Needed to Cope with the Impact of Tourists

Tourists cause traffic congestion, erode footpaths and drop litter.
Here are a few strategies being carried out to reduce these problems:

1) **Coping with the extra traffic** — public transport in the area is being improved so people can leave their cars at home. There are also campaigns to encourage people to use the new services, e.g. the 'Give the driver a break' campaign. This provides leaflets that show the routes available and offers discounts at cafes and on lake cruises for people presenting bus or train tickets.

2) **Coping with the erosion of footpaths** — solutions include encouraging visitors to use less vulnerable areas instead, 'resting' popular routes by changing the line of the paths, and using more hard-wearing materials for paths. E.g. at Tarn Hows, severely eroded paths have been covered with soil and reseeded, and the main route has been gravelled to protect it.

3) **Protecting wildlife and farmland** — there are signs to remind visitors to take their litter home and covered bins are provided at the most popular sites. There have also been campaigns to encourage visitors to enjoy the countryside responsibly, e.g. by closing gates and keeping dogs on a lead.

There are Plans to Make Sure it Keeps Attracting Tourists

1) The official tourism strategy for Cumbria is to attract an extra two million visitors by 2018 and to increase the amount tourists spend from £1.1 billion per year to £1.5 billion per year.
2) Public transport will be improved to make the Lakes even more accessible.
3) There's to be widespread advertising and marketing to make the area even more well known.
4) Farms will be encouraged to provide services like quad biking, clay pigeon shooting and archery alongside traditional farming — these should attract more tourists to the area.
5) Timeshare developments (where people share the ownership of a property, but stay there at different times) are to be increased, to help bring people into the area all year round.
6) The strategy also aims to encourage tourism in areas outside the National Park, like the West Coast, Furness and Carlisle, to relieve some of the pressure on the main tourist areas. E.g. there are plans to regenerate ports like Whitehaven and Barrow to make them more attractive to visitors.

Not much chance of wandering lonely as a cloud in the Lakes nowadays...

Trust me, visiting the Lakes is a lot more enjoyable than learning about the area's strategies for coping with tourism — but the learning part's all you're going to get today (unless you're reading this whilst on a field trip to the Lakes, you lucky devil).

Mass Tourism

For me, 'mass tourism' conjures up images of pasty Brits swigging lager in Spanish coastal resorts, but there's a bit more to it than that. It can have a big impact on the areas the tourists flock to.

Mass Tourism is Basically Tourism on a Big Scale

Mass tourism is organised tourism for large numbers of people.
For example, visiting Spain on a package holiday would count as mass tourism.
But, holidays where people organise it themselves or small group tours don't count.

Mass Tourism has Both Positive and Negative Impacts

	POSITIVE	NEGATIVE
ECONOMIC IMPACTS	• It brings money into the local economy. • It creates jobs for local people, and increases the income of industries that supply tourism, e.g. farming.	• A lot of the profit made from tourism is kept by the large travel companies, rather than going to the local economy.
SOCIAL IMPACTS	• Lots of jobs means young people are more likely to stay in the area. • Improved roads, communications and infrastructure for tourists also benefit local people. • Income from tourism can be reinvested in local community projects.	• The tourism jobs available to locals are often badly paid and seasonal. • Traffic congestion caused by tourists can inconvenience local people. • The behaviour of some tourists can offend locals.
ENVIRONMENTAL IMPACTS	• Income from tourism can be reinvested in protecting the environment, e.g. to run National Parks or pay for conservation work.	• Transporting lots of people long distances releases lots of greenhouse gases that cause global warming. • Tourism can increase litter and cause pollution, e.g. increased sewage can cause river pollution. • Tourism can lead to the destruction of natural habitats, e.g. sightseeing boats can damage coral reefs.

There are Ways to Reduce the Negative Impacts of Mass Tourism

Here are a few examples:

1) Improving public transport encourages tourists to use it, which reduces congestion and pollution.
2) Limiting the number of people visiting sensitive environments, e.g. coral reefs, reduces damage.
3) Providing lots of bins helps to reduce litter.

The Importance of Tourism Needs to be Maintained

Areas that rely heavily on tourism need to make sure the tourists keep coming.
Here are a few ways they can do this:

1) Build new facilities or improve existing ones, e.g. build new hotels.
2) Reduce any tourist impacts that make the area less attractive, e.g. litter and traffic congestion.
3) Advertise and market the area to attract new tourists, e.g. use TV to advertise in other countries.
4) Improve transport infrastructure to make it quicker and easier to get to the area.
5) Offer new activities to attract tourists that don't normally go there.
6) Make it cheaper to visit, e.g. lower entrance fees to attractions.

O' I do like to live beside the seaside — or I did before that coach party turned up...

Pros and cons, that's what this page is about — the pros and cons of mass tourism obviously. Nothing to do with the criminal fraternity. There's also a bit about how tourism should be managed to stop those pesky cons from taking over.

Mass Tourism — Case Study

If watching lions devour a gazelle while on holiday is your bag, then Kenya is the place to go.

Kenya is a Popular Tourist Destination

Kenya is in East Africa. It gets over 700 000 visitors per year. There are a few reasons why people visit:

1) A fascinating tribal culture and lots of wildlife, including the 'big five' (rhino, lion, elephant, buffalo and leopard). Wildlife safaris are very popular.
2) A warm climate with sunshine all year round.
3) Beautiful scenery, including savannah, mountains, forests, beaches and coral reefs.

Tourism has Had a Big Impact on Kenya

	POSITIVE	NEGATIVE
ECONOMIC IMPACTS	• Tourism contributes 15% of the country's Gross National Product. • In 2003, around 219 000 people worked in the tourist industry.	• Only 15% of the money earned through tourism goes to locals. The rest goes to big companies.
SOCIAL IMPACTS	• The culture and customs of the native Maasai tribe are preserved because things like traditional dancing are often displayed for tourists.	• Some Maasai tribespeople were forced off their land to create National Parks for tourists. • Some Muslim people in Kenya are offended by the way female tourists dress.
ENVIRONMENTAL IMPACTS	• There are 23 National Parks in Kenya, e.g. Nairobi National Park. Tourists have to pay entry fees to get in. This money is used to maintain the National Parks, which help protect the environment and wildlife.	• Safari vehicles have destroyed vegetation and caused soil erosion. • Wild animals have been affected, e.g. cheetahs in the most heavily visited areas have changed their hunting behaviour to avoid the crowds. • Coral reefs in the Malindi Marine National Park have been damaged by tourist boats anchoring.

Kenya is Trying to Reduce the Negative Impacts of Tourism

1) Walking or horseback tours are being promoted over vehicle safaris, to preserve vegetation.
2) Alternative activities that are less damaging than safaris are also being encouraged, e.g. climbing and white water rafting.

Kenya is Also Trying to Maintain Tourism

1) Kenya's Tourist Board and Ministry of Tourism have launched an advertising campaign in Russia called 'Magical Kenya'.
2) Kenya Wildlife Service is planning to build airstrips in Ruma National Park and Mount Elgon National Park to make them more accessible for tourists. It also plans to spend £8 million improving roads, bridges and airstrips to improve accessibility.
3) Visa fees for adults were cut by 50% in 2009 to make it cheaper to visit the country. They were also scrapped for children under 16 to encourage more families to visit.

Shh! It's the lesser-spotted tourist in its natural habitat — the poolside bar...

A walking safari in Kenya sounds dodgy — lions eat people and elephants can be pretty stroppy when they want to be. It's just one way of reducing the impacts of tourism though (which reminds me — check you know the impacts too).

Tourism in Extreme Environments

Some people aren't content with a week in the sun or a shopping spree in New York — they go on holiday to extreme environments, e.g. Antarctica, the Himalayas and the Sahara desert.

Extreme Environments are Becoming Popular with Tourists

There are many reasons why tourists are attracted to extreme environments:

1) They're ideal settings for adventure holiday activities like jeep tours, river rafting and trekking.
2) Some people want something different and exciting to do on holiday, which nobody else they know has done.

3) A lot of people enjoy an element of risk and danger in their leisure time, which the harsh conditions of an extreme environment can provide.
4) Some wildlife can only be seen in these areas, e.g. polar bears can only be seen in the Arctic.
5) Some scenery can only be seen in extreme places too, e.g. icebergs can only be seen in very cold environments.

There are also several reasons why tourism is increasing in extreme environments:

1) Improvements in transport have made it quicker and easier to get to some of these destinations. For example, the Qinghai-Tibet railway that links China and Tibet (an extreme mountain environment) opened in 2006. This increased tourism as Tibet was easier to get to.
2) People are keen to see places like Antarctica for themselves while they have the chance, before the ice melts due to global warming.
3) Tourism to extreme environments is quite expensive, but people nowadays tend to have more disposable income (spare cash), so more people can afford to go.
4) Adventure holidays are becoming more popular because of TV programmes and advertising.

Tourism in Extreme Environments can be Damaging

The ecosystems in extreme environments are usually delicately balanced, because it's so difficult for life to survive in the harsh conditions there. The presence of tourists can upset this fragile balance and cause serious problems. Here's an example of how tourism can damage the environment in the Himalayas:

1) Trees are cut down to provide fuel for trekkers and other tourists, leading to deforestation.
2) Deforestation destroys habitats.
3) Deforestation also means there are fewer trees to intercept rain. So more water reaches channels causing flooding.
4) Tree roots normally hold the soil together, so deforestation also leads to soil erosion. If soil is washed into rivers it raises the river bed so it can't hold as much water — this can cause flooding too.
5) The sheer volume of tourists causes footpath erosion, which can lead to landslides.
6) Toilets are poor or non-existent, so rivers become polluted by sewage.

The inside of my fridge — now that's a pretty extreme environment...

You might be asked to suggest reasons why people go to extreme environments, and answering 'because they're mad' won't cut it. So check you've got the reasons covered and have an extreme case study ready for 'em too.

Tourism in Extreme Environments — Case Study

Antarctica is the coldest place on Earth (it can get to minus 80 °C), making it an extreme environment. Despite this fact quite a few tourists don their thermal knickers and brave the cold every year.

The Antarctic is Becoming More Popular with Tourists

1) Antarctica is a continent at the Earth's South Pole. It covers an area of about 14 million km^2 and about 98% is covered with ice.
2) The number of tourists visiting Antarctica each year is rising, e.g. there were 7413 in the 1996/1997 season, but 46 000 in the 2007/2008 season.
3) Tourists are attracted by the stunning scenery (e.g. icebergs) and the wildlife (e.g. penguins and whales).

Tourism has Environmental Impacts in Antarctica

Antarctica is very cold and doesn't get much sunshine in winter so the land ecosystems are very fragile — it takes a long time for them to recover from damage. The sea ecosystem is also delicately balanced. This means that tourists can have a massive impact on the environment there:

1) Tourists can trample plants, disturb wildlife and drop litter.
2) There are fears that tourists could accidentally introduce non-native species or diseases that could wipe out existing species.
3) Spillage of fuel from ships is also a worry, especially after the sinking of the cruise ship, MS Explorer, in 2007. Fuel spills kill molluscs (e.g. mussels) and fish, as well as the birds that feed on them (e.g. penguins).

There are Measures in Place to Protect Antarctica

(1) The Antarctic Treaty is an international agreement that came into force in 1961 and has now been signed by 47 countries. The Treaty is designed to protect and conserve the area and its plant and animal life. In April 2009, the parties involved with the Antarctic Treaty agreed to introduce new limits on tourism in Antarctica — only ships with fewer than 500 passengers are allowed to land there and a maximum of 100 passengers are allowed on shore at a time.

(2) The International Association of Antarctica Tour Operators also has a separate Code of Conduct. The code is voluntary, but most operators in the area do stick to it. There are rules on:

1) Specially Protected Areas — these are off limits to tourists.
2) Wildlife — wildlife must not be disturbed when being observed. E.g. when whale watching, boats should approach animals slowly and keep their distance.
3) Litter — nothing can be left behind by tourists and there must be no smoking during shore landings (to reduce cigarette end litter).
4) Supervision — tourists must stay with their group and each group must have a qualified guide. This prevents people from entering no-go areas or disturbing wildlife.
5) Plant life — tourists must not walk on the fragile plant life.
6) Waste — sewage must be treated biologically and other waste stored on board the ships.

Please keep your voices down — the penguins are having an off day...

Time to get your 'case study hat' on — I realise you might have been wearing it quite a lot throughout this section, but hey, we're on holiday. And oh, I wouldn't throw it away when you've finished this page either. Just a hint.

Ecotourism

As if UK tourism, mass tourism and extreme tourism weren't enough — it's time for ecotourism...

Ecotourism Doesn't Destroy the Environment

1) Ecotourism is tourism that doesn't harm the environment and benefits the local people.
2) Ecotourism involves:
   - Conservation — protecting and managing the environment.
   - Stewardship — taking responsibility for conserving the environment.
3) Ideally, conservation and stewardship should involve local people and local organisations, so that local people benefit from the tourists.
4) Ecotourism is usually a small-scale activity, with only small numbers of visitors going to an area at a time. This helps to keep the environmental impact of tourism low.
5) It often involves activities like wildlife viewing and walking.

Ecotourism Benefits the Environment, Economy and Local People

Environmental benefits:

1) Local people are encouraged to conserve the environment rather than use it for activities that can be damaging, e.g. logging or farming. This is because they can only earn money from ecotourism if the environment isn't damaged.
2) It reduces poaching and hunting of endangered species, since locals will benefit more from protecting these species for tourism than if they killed them.
3) Ecotourism projects try to reduce the use of fossil fuels, e.g. by using renewable energy sources and local food (which isn't transported as far so less fossil fuel is used). Using less fossil fuel is better for the environment as burning fossil fuels adds to global warming.
4) Waste that tourists create is disposed of carefully to prevent pollution.

Economic benefits:

1) Ecotourism creates jobs for local people (e.g. as guides or in tourist lodges), which helps the local economy grow.
2) Local people not directly employed in tourism can also make money by selling local crafts to visitors or supplying the tourist industry with goods, e.g. food.

Benefits for local people:

1) People have better and more stable incomes in ecotourism than in other jobs, e.g. farming.
2) Many ecotourism schemes fund community projects, e.g. schools, water tanks and health centres.

Ecotourism Helps the Sustainable Development of Areas

1) Sustainable development means improving the quality of life for people, but doing it in a way that doesn't stop people in the future getting what they need (by not damaging the environment or depleting resources).
2) Ecotourism helps areas to develop by increasing the quality of life for local people — the profits from ecotourism can be used to build schools or healthcare facilities.
3) The development is sustainable because it's done without damaging the environment — without ecotourism people may have to make a living to improve their lives by doing something that harms the environment, e.g. cutting down trees.

Not to be confused with egotourism — holidays on board a big golden yacht...

Ecotourism is sustainable because it's something that can continue into the future — the area remains unspoilt, so tourists will continue to come and enjoy it. It helps give local people a better quality of life too.

Ecotourism — Case Study

I know you're a bit sick of case studies by now, but this is the last one in the section — scout's promise.

Tataquara Lodge is an Example of Ecotourism

1) Tataquara Lodge is on an island in the Xingu River in the Brazilian state of Para.
2) It's owned and operated by a cooperative of six local tribes of indigenous people.
3) The lodge has 15 rooms and offers activities like fishing, canoeing, wildlife viewing and forest walks.
4) The surrounding rainforest is home to a rich variety of wildlife, including many species of bat and tropical birds. There are also some endangered species in the area, such as the harpy eagle and giant river otter.

The Lodge Has Many Benefits

Environmental benefits:

- The lodge was built from local materials such as straw and wood that was found on the ground — this means they didn't have to cut down any trees. These materials also make the buildings blend in with the natural environment so they don't spoil the scenery.
- It uses solar power to run lights, rather than burning fossil fuels to generate electricity, which is better for the environment.
- The food served in the lodge is all locally-produced. This means less fossil fuel is used to transport it than if it came from further away.

Economic benefits:

- The lodge is owned by a cooperative of indigenous tribes rather than a big foreign company, so the income it provides goes straight to the local economy.
- As the lodge uses locally-produced food, more money goes back into the local economy.

Benefits for local people:

- The lodge creates jobs for local people.
- People in nearby villages are encouraged to visit Tataquara Lodge to sell crafts and perform traditional songs and dances — this gives them an income and helps preserve their culture.
- Profits earned from the lodge are used to provide decent healthcare and education for thousands of people from the local tribes.

Tataquara Lodge Helps the Sustainable Development of the Area

1) Profits from Tataquara lodge are used to improve healthcare and education in the area. This helps the area to develop by increasing the quality of life for the local people.
2) The development is sustainable because the money to do it is generated without damaging the environment — local people don't have to find other employment that could damage the environment, e.g. logging or farming. Also, resources aren't used up, e.g. solar power is used to run lights instead of fossil fuels, so more resources are available for future generations.

My local cooperative does a lovely strawberry cheesecake...

Tataquara is a bit of a mouthful to say, but it's a great example of how ecotourism benefits the environment and the local people. Check you can remember plenty of facts about it in case you get asked a question on ecotourism.

Revision Summary for Section 14

It's that time again — just when you think you're all done and dusted with a section, a page of questions is sprung on you. This lot are all about going on your holidays though, so they shouldn't be too much of a strain. Give them a try, and then if there are any you struggle with you can go back through the section and pick up even more ideas for your next trip.

1) Give three reasons why global tourism is increasing.
2) Why do cities attract large numbers of tourists?
3) What attracts tourists to mountain areas?
4) Why is tourism important to the economies of many countries?
5) How much money does the tourist industry contribute to the UK's economy?
6) Give two factors that can decrease the number of tourists visiting the UK.
7) Describe the six stages in the tourist area life cycle model.
8) a) Name an area in the UK that's popular with tourists.
   b) Describe what attracts tourists to the area.
   c) Describe the strategies used to reduce the impact of tourists.
   d) How does this area plan to keep attracting tourists in the future?
9) Define mass tourism.
10) List three positive effects and three negative effects of mass tourism.
11) Give three examples of how mass tourism can be managed to ensure that an area keeps its appeal.
12) a) Name an area that has mass tourism.
   b) Describe what makes this area such a popular tourist destination.
   c) Describe the negative impacts tourism has had on the area.
   d) What is being done to help reduce the negative impacts?
13) Give three reasons why tourism in extreme environments is increasing.
14) a) Name an extreme environment that is becoming popular with tourists.
   b) Describe the environmental impacts of tourism in the area.
   c) Describe the measures in place to limit the impact of tourism in the area.
15) Define ecotourism.
16) Explain one way that ecotourism can benefit the economy of a region.
17) Explain one way that ecotourism can benefit the environment of a region.
18) How can ecotourism help benefit local people?
19) a) What is sustainable development?
   b) Explain how ecotourism helps the sustainable development of areas.
20) a) Give an example of a successful ecotourism project.
   b) Explain how this project benefits the local environment, the local economy and the local people.

Answering Questions

This section is filled with lots of lovely techniques and skills that will be useful in your exam. It's no good learning the content of this book if you don't bother learning the skills that will help you to pass your exam too. First up, answering questions properly...

Make Sure you Read the Question Properly

It's dead easy to misread the question and spend five minutes writing about the wrong thing.
Five simple tips can help you avoid this:

1) Figure out if it's a case study question — if the question says something like 'using named examples' or 'with reference to one named area' you need to include a case study.
2) Underline the command words in the question (the ones that tell you what to do):

Answers to questions with 'explain' in them often include the word 'because' (or 'due to').

When writing about differences, 'whereas' is a good word to use in your answers, e.g. 'the Richter scale measures the energy released by an earthquake whereas the Mercalli scale measures the effects'.

Command word	Means write about...
Describe	what it's like
Explain	why it's like that (i.e. give reasons)
Compare	the similarities AND differences
Contrast	the differences
Suggest why	give reasons for

If a question asks you to describe a pattern (e.g. from a map or graph), make sure you identify the general pattern, then refer to any anomalies (things that don't fit the general pattern).

E.g. to answer 'describe the global distribution of volcanoes', first say that they're mostly on plate margins, then mention that a few aren't (e.g. in Hawaii).

3) Underline the key words (the ones that tell you what it's about), e.g. volcanoes, erosion, migration, rural-urban fringe, population pyramid.
4) If the question says 'Use evidence from Fig. 2...' you need to refer to the figure in your answer, e.g. quote numbers from it to back up your points.
5) Re-read the question and your answer when you've finished, just to check that what you've written really does answer the question being asked. A common mistake is to miss a bit out — like when questions say 'use data from the graph in your answer' or 'use evidence from the map'.

Case Study Questions are Level Marked

Case study questions are often worth 8 marks and are level marked, which means you need to do these things to get the top level (3) and a high mark:

1) Read the question properly and figure out a structure before you start. Your answer needs to be well organised and structured, and written in a logical way.
2) Include specialist terms (geographical words), e.g. destructive margin, backwash, longshore drift.
3) Include plenty of relevant details:
    - This includes things like names, dates, statistics, names of organisations or companies.
    - Don't forget that they need to be relevant though — it's no good including the exact number of people killed in a flood when the question is about the causes of a flood.
4) Your answer should be legible (you won't get many marks if the examiner can't read it).
5) There will often be extra marks available for spelling, punctuation and grammar — so make sure you write carefully and check your work through.

Describe the similarities and differences between compare and contrast...

It may all seem a bit simple to you, but it's really important to understand what you're being asked to do. This can be tricky — sometimes the differences between the meanings of the command words are quite subtle.

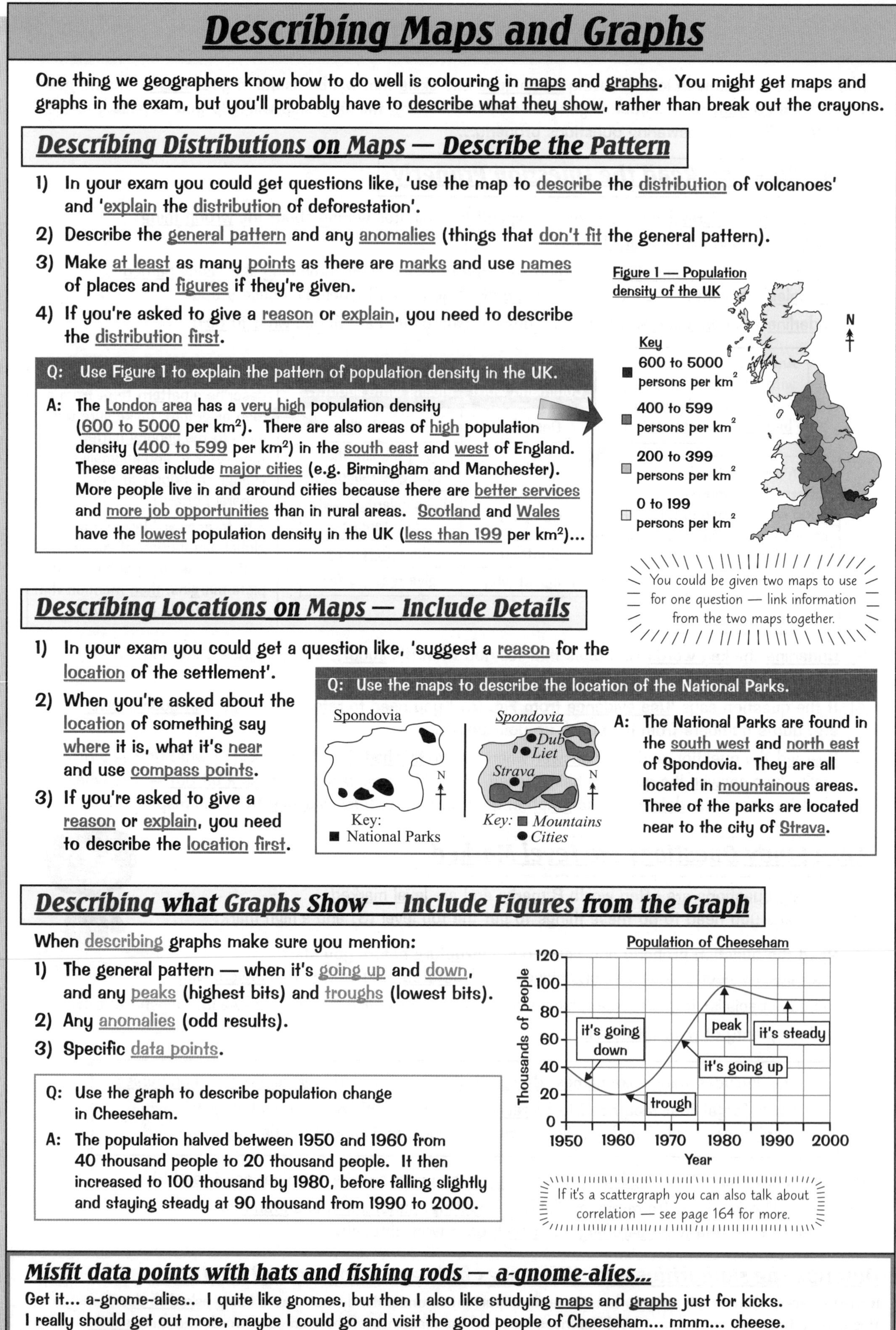

Describing Maps and Graphs

One thing we geographers know how to do well is colouring in maps and graphs. You might get maps and graphs in the exam, but you'll probably have to describe what they show, rather than break out the crayons.

Describing Distributions on Maps — Describe the Pattern

1) In your exam you could get questions like, 'use the map to describe the distribution of volcanoes' and 'explain the distribution of deforestation'.
2) Describe the general pattern and any anomalies (things that don't fit the general pattern).
3) Make at least as many points as there are marks and use names of places and figures if they're given.
4) If you're asked to give a reason or explain, you need to describe the distribution first.

Q: Use Figure 1 to explain the pattern of population density in the UK.

A: The London area has a very high population density (600 to 5000 per km^2). There are also areas of high population density (400 to 599 per km^2) in the south east and west of England. These areas include major cities (e.g. Birmingham and Manchester). More people live in and around cities because there are better services and more job opportunities than in rural areas. Scotland and Wales have the lowest population density in the UK (less than 199 per km^2)...

Figure 1 — Population density of the UK

You could be given two maps to use for one question — link information from the two maps together.

Describing Locations on Maps — Include Details

1) In your exam you could get a question like, 'suggest a reason for the location of the settlement'.
2) When you're asked about the location of something say where it is, what it's near and use compass points.
3) If you're asked to give a reason or explain, you need to describe the location first.

Q: Use the maps to describe the location of the National Parks.

A: The National Parks are found in the south west and north east of Spondovia. They are all located in mountainous areas. Three of the parks are located near to the city of Strava.

Describing what Graphs Show — Include Figures from the Graph

When describing graphs make sure you mention:

1) The general pattern — when it's going up and down, and any peaks (highest bits) and troughs (lowest bits).
2) Any anomalies (odd results).
3) Specific data points.

Q: Use the graph to describe population change in Cheeseham.

A: The population halved between 1950 and 1960 from 40 thousand people to 20 thousand people. It then increased to 100 thousand by 1980, before falling slightly and staying steady at 90 thousand from 1990 to 2000.

Population of Cheeseham

If it's a scattergraph you can also talk about correlation — see page 164 for more.

Misfit data points with hats and fishing rods — a-gnome-alies...

Get it... a-gnome-alies.. I quite like gnomes, but then I also like studying maps and graphs just for kicks. I really should get out more, maybe I could go and visit the good people of Cheeseham... mmm... cheese.

Charts and Graphs

The next four pages are filled with lots of different types of charts, graphs and maps. There are two important things to learn — NUMBER ONE: how to interpret them (read them), and NUMBER TWO: how to construct and complete them (fill them in). You might have to do it in the exam so pay attention.

Bar Charts — Draw the Bars Straight and Neat

1) Reading Bar Charts

1) Read along the bottom to find the bar you want.
2) To find out the value of a bar in a normal bar chart — go from the top of the bar across to the scale, and read off the number.
3) To find out the value of part of the bar in a divided bar chart — find the number at the top of the part of the bar you're interested in, and take away the number at the bottom of it.

Q: How many barrels of oil did Oxo oil produce per day in 2008?

A: 500 thousand – 350 thousand = 150 thousand barrels per day

Oil production

2007
2008
Line across from 350
Thousands of barrels per day
0
100
200
300
400
500
Oxo oil
Gnoxo Ltd.
Froxo Inc.
Company

2) Completing Bar Charts

1) First find the number you want on the vertical scale.
2) Then trace a line across to where you want the top of the bar to be with a ruler.
3) Draw in a bar of the right size using a ruler.

Q: Complete the chart to show that Froxo Inc. produced 200 thousand barrels of oil per day in 2008.

A: 150 thousand (2007) + 200 thousand = 350 thousand barrels. So draw the bar up to this point.

Line Graphs — the Points are Joined by Lines

1) Reading Line Graphs

1) Read along the correct scale to find the value you want, e.g. 20 thousand tonnes or 1920.
2) Read across or up to the line you want, then read the value off the other scale.

Q: How much coal did New Wales Ltd. produce in 1900?

A: Find 1900 on the bottom scale, go up to the red line, read across, and it's 20 on the scale. The scale's in thousands of tonnes, so the answer is 20 thousand tonnes.

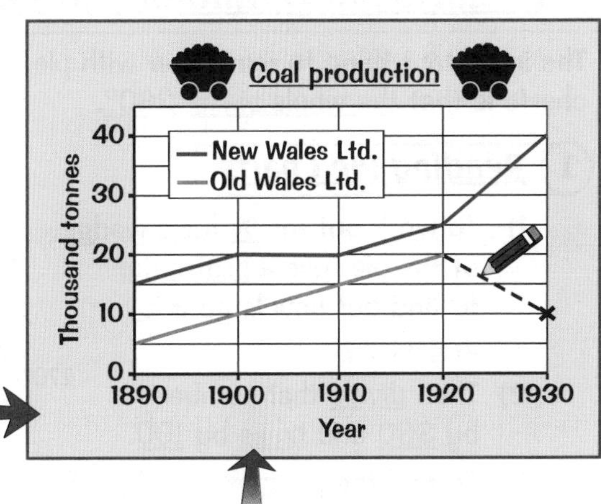

2) Completing Line Graphs

1) Find the value you want on both scales.
2) Make a mark (e.g. ×) at the point where the two values meet on the graph.
3) Using a ruler, join the mark you've made to the line that it should be connected to.

Q: Complete the graph to show that Old Wales Ltd. produced 10 thousand tonnes of coal in 1930.

A: Find 1930 on the bottom scale, and 10 thousand tonnes on the vertical scale. Make a mark where they meet, then join it to the blue line with a ruler.

The top forty for sheep — the baaaaaaaaaaaa chart...

Something to watch out for with bar charts and line graphs is reading the scale — check how much each division is worth before reading them or completing them. It's easy to think they're always worth one each, but sadly not.

Charts and Graphs

'More charts and graphs' I hear you cry — well OK, your weird wishes are my command.

Scatter Graphs Show Relationships

Scatter graphs tell you how closely related two things are, e.g. rainfall and river discharge. The fancy word for this is correlation. Strong correlation means the two things are closely related to each other. Weak correlation means they're not very closely related. The line of best fit is a line that goes roughly through the middle of the scatter of points and tells you about what type of correlation there is. Data can show three types of correlation:

1) Positive — as one thing increases the other increases.
2) Negative — as one thing increases the other decreases.
3) None — there's no relationship between the two things.

Line of best fit

Positive Negative None

1) Reading Scatter Graphs

1) If you're asked to describe the relationship, look at the slope of the graph, e.g. if the line's moving upwards to the right it's a positive correlation. You also need to look at how close the points are to the line of best fit — the closer they are the stronger the correlation.
2) If you're asked to read off a specific point, just follow the rules for a line graph (see previous page).

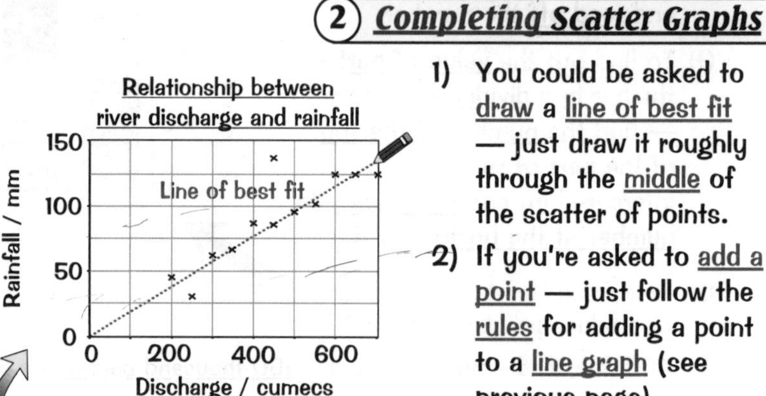

Q: Describe the relationship shown by the scatter graph.

A: River discharge and rainfall show a strong, positive correlation — as rainfall increases, so does river discharge.

2) Completing Scatter Graphs

1) You could be asked to draw a line of best fit — just draw it roughly through the middle of the scatter of points.
2) If you're asked to add a point — just follow the rules for adding a point to a line graph (see previous page).

Pie Charts Show Amounts or Percentages

The important thing to remember with pie charts is that the whole pie = 360°.

1) Reading Pie Charts

1) To work out the % for a wedge of the pie, use a protractor to find out how large it is in degrees.
2) Then divide that number by 360 and times by 100.
3) To find the amount a wedge of the pie is worth, work out your percentage then turn it into a decimal. Then times the decimal by the total amount of the pie.

Pie Chart of Transport Type

0° 324° 90° 270° 126° 180° Bicycle Pogostick Car

Q: Out of 100 people, how many used a pogostick?

A: 126 – 90 = 36°, so (36 ÷ 360) × 100 = 10%, so 0.1 × 100 = 10 people.

2) Completing Pie Charts

1) To draw on a new wedge that you know the % for, turn the % into a decimal and times it by 360. Then draw a wedge of that many degrees.

Q: Out of 100 people, 25% used a bicycle. Add this to the pie chart.

A: 25 ÷ 100 = 0.25, 0.25 × 360 = 90°.

2) To add a new wedge that you know the amount for, divide your amount by the total amount of the pie and times the answer by 360. Then draw on a wedge of that many degrees.

Q: Out of 100 people, 55 used a car, add this to the pie chart.

A: 55 ÷ 100 = 0.55, 0.55 × 360 = 198° (198° + 126° = 324°).

Sorry darling, we've got no relationship — look at our scatter graph...

Hmm, who'd have thought pie could be so complicated. Don't panic though, a bit of practice and you'll be fine. And don't worry, you're over half way through this section now. Congratulations — I'm so proud of you, sniff.

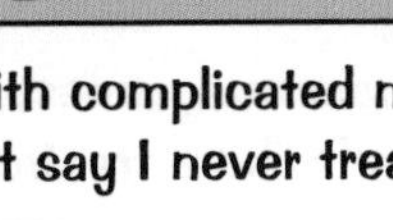

Maps

A couple of jazzy maps on this page for you, both with complicated names — topological and proportional symbol. And a bit on isolines too. Don't say I never treat you...

Topological Maps are Simplified Maps

1) Some maps are hard to read because they show too much detail.
2) Topological maps get around this by just showing the most important features like roads and rail lines. They don't have correct distances or directions either, which makes them easier to read.
3) They're often used to show transport networks, e.g. the London tube map.
4) If you have to read a topological map — dots are usually places and lines usually show routes between places. If two lines cross at a dot then it's usually a place where you can switch routes.
5) As always, don't forget to check out the key.

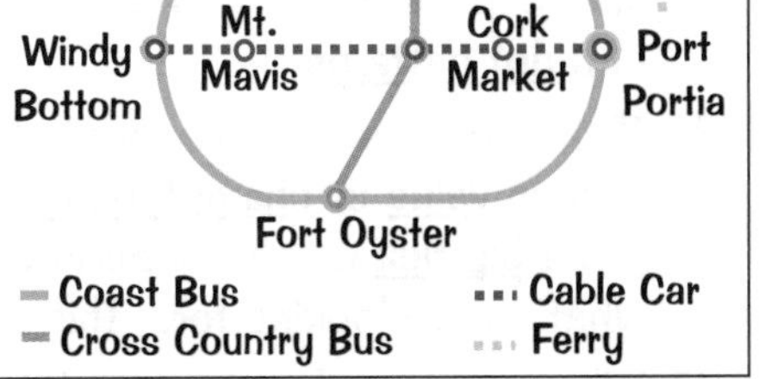

Q: How many different transport routes pass through Port Portia?

A: Three (bus, cable car and ferry).

Proportional Symbol Maps use Symbols of Different Sizes

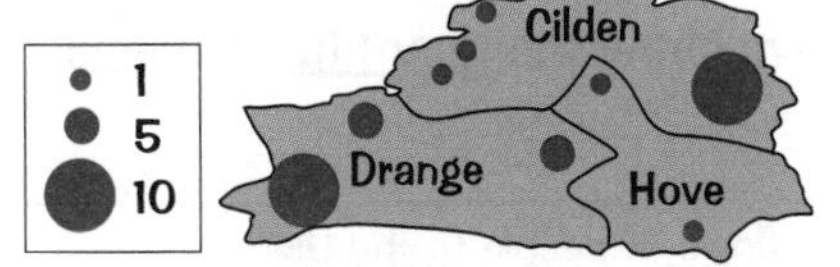

Q: Which area of Drumshire has the most car parks?

A: Drange, with 20.

1) Proportional symbol maps use symbols of different sizes to represent different quantities.
2) A key shows the quantity each different sized symbol represents. The bigger the symbol, the larger the amount.
3) The symbols might be circles, squares, semi-circles or bars, but a larger symbol always means a larger amount.

Isolines on Maps Link up Places with Something in Common

1) Isolines are lines on a map linking up all the places where something's the same, for example:
   - Contour lines are isolines linking up places at the same altitude (see p. 168).
   - Isolines on a weather map (called isobars) link together all the places where the pressure's the same.
2) Isolines can be used to link up lots of things, e.g. average temperature, wind speed or rainfall.

1) Reading Isoline Maps

1) Find the place you're interested in on the map and if it's on a line just read off the value.
2) If it's between two lines, you have to estimate the value.

Q: Find the average annual rainfall in Port Portia and on Mt. Mavis.

A: Port Portia is about half way between the lines for 200 mm and 400 mm so the rainfall is around 300 mm per year. Mt. Mavis is on an isoline so the rainfall is 1000 mm per year.

Average annual rainfall on Itchy Island (mm per year)

N
600
500
600
1000
500
Mt. Mavis
800
Port Portia
600
400
200

2) Completing Isoline Maps

1) Drawing an isoline's like doing a dot-to-dot — you just join up all the dots with the same numbers.
2) Make sure you don't cross any other isolines though.

Q: Complete on the map the isoline showing an average rainfall of 600 mm per year.

A: See the red line on the map.

Lines and lines and lines...

If you have to draw an isoline on a map, then check all the info on the map before you start drawing. If you know where the line's got to go you won't muck it up. Make sure you do it in pencil too, so you can rub out any mistakes.

Maps

Three more maps, with three more ludicrous names. Well the last two aren't that bad, but this first one — choropleth, sounds like a treatment at the dentist.

Choropleth Maps show How Something Varies Between Different Areas

1) Choropleth maps show how something varies between different areas using colours or patterns.
2) The maps in exams often use cross-hatched lines and dot patterns.
3) If you're asked to talk about all the parts of the map with a certain value or characteristic, look at the map carefully and put a big tick on all the parts with the pattern that matches what you're looking for. This makes them all stand out.
4) When you're asked to complete part of a map, first use the key to work out what type of pattern you need. Then carefully draw on the pattern, e.g. using a ruler.

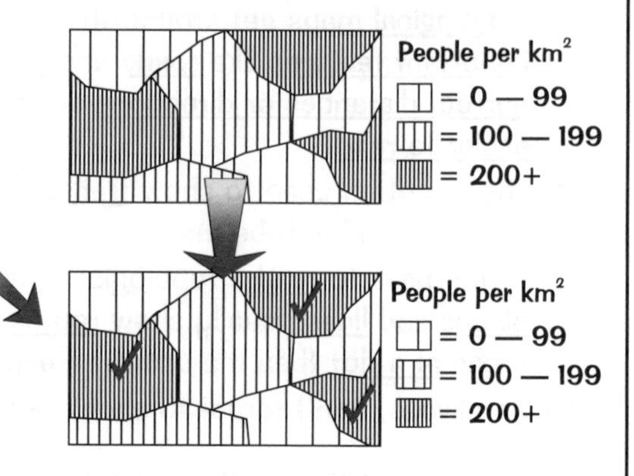

Flow Lines show Movement

1) Flow line maps have arrows on, showing how things move (or are moved) from one place to another.
2) They can also be proportional symbol maps — the width of the arrows show the quantity of things that are moving.

Q: From which area do the greatest number of people entering the UK come from?

A: USA, as this arrow is the largest.

Q: The number of people entering the UK from the Middle East is roughly half the number of people entering from the USA. Draw an arrow on the map to show this.

A: Make sure your arrow is going in the right direction and its size is appropriate (e.g. half the width of the USA arrow).

Some of the flows of people to the UK

USA

Middle East

Rest of the Americas

Immigration

Desire Lines show Journeys

1) Desire line maps are a type of flow line as they show movement too.
2) They're straight lines that show journeys between two locations, but they don't follow roads or railway lines.
3) One line represents one journey.
4) They're used to show how far all the people have travelled to get to a place, e.g. a shop or a town centre, and where they've come from.

Desire Lines showing journeys to Cheeseham

Desire lines — I'm sure my palm reader mentioned those...

...unfortunately I'm not as good at seeing the future as she is* so I can't predict if any of these maps are going to come up in your exam. They could do though, so check you know what they are and how to read them.

*If you're wondering, I'm going to meet a tall, dark, handsome stranger very soon...

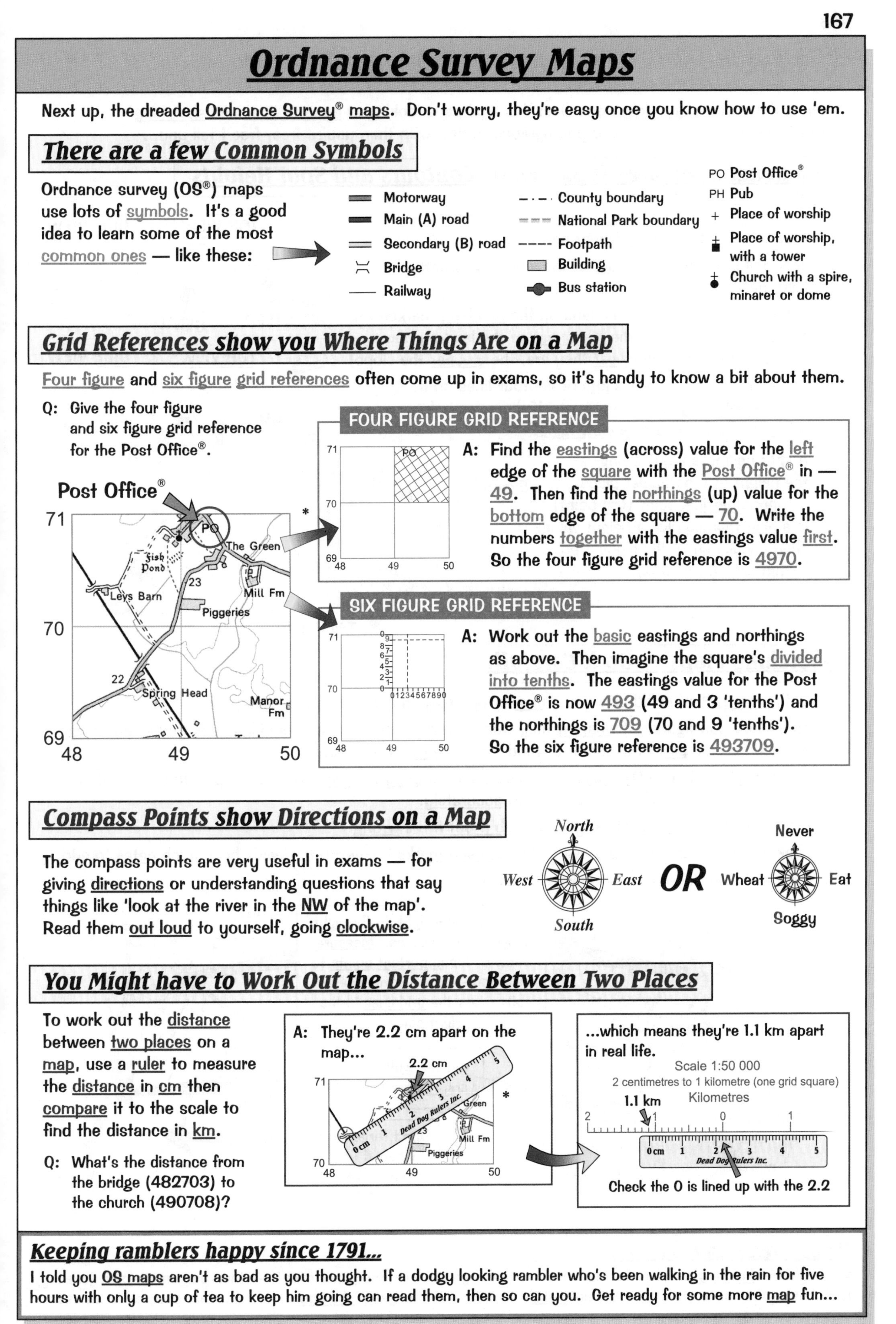

Ordnance Survey Maps

Next up, the dreaded Ordnance Survey® maps. Don't worry, they're easy once you know how to use 'em.

There are a few Common Symbols

Ordnance survey (OS®) maps use lots of symbols. It's a good idea to learn some of the most common ones — like these:

Grid References show you Where Things Are on a Map

Four figure and six figure grid references often come up in exams, so it's handy to know a bit about them.

Q: Give the four figure and six figure grid reference for the Post Office®.

FOUR FIGURE GRID REFERENCE

A: Find the eastings (across) value for the left edge of the square with the Post Office® in — 49. Then find the northings (up) value for the bottom edge of the square — 70. Write the numbers together with the eastings value first. So the four figure grid reference is 4970.

SIX FIGURE GRID REFERENCE

A: Work out the basic eastings and northings as above. Then imagine the square's divided into tenths. The eastings value for the Post Office® is now 493 (49 and 3 'tenths') and the northings is 709 (70 and 9 'tenths'). So the six figure reference is 493709.

Compass Points show Directions on a Map

The compass points are very useful in exams — for giving directions or understanding questions that say things like 'look at the river in the NW of the map'. Read them out loud to yourself, going clockwise.

You Might have to Work Out the Distance Between Two Places

To work out the distance between two places on a map, use a ruler to measure the distance in cm then compare it to the scale to find the distance in km.

Q: What's the distance from the bridge (482703) to the church (490708)?

A: They're 2.2 cm apart on the map...

...which means they're 1.1 km apart in real life.

Check the 0 is lined up with the 2.2

Keeping ramblers happy since 1791...

I told you OS maps aren't as bad as you thought. If a dodgy looking rambler who's been walking in the rain for five hours with only a cup of tea to keep him going can read them, then so can you. Get ready for some more map fun...

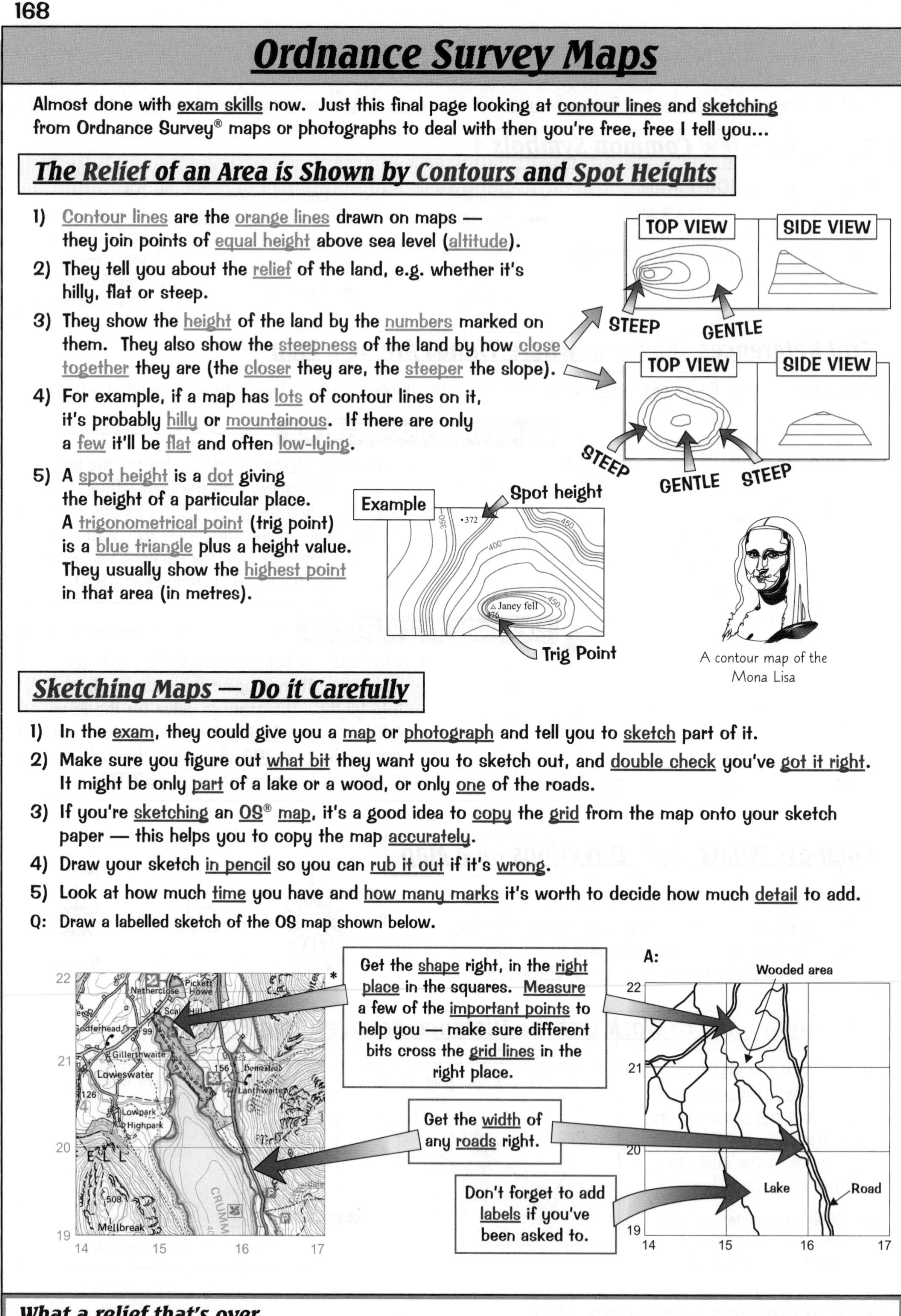

Ordnance Survey Maps

Almost done with exam skills now. Just this final page looking at contour lines and sketching from Ordnance Survey® maps or photographs to deal with then you're free, free I tell you...

The Relief of an Area is Shown by Contours and Spot Heights

1) Contour lines are the orange lines drawn on maps — they join points of equal height above sea level (altitude).
2) They tell you about the relief of the land, e.g. whether it's hilly, flat or steep.
3) They show the height of the land by the numbers marked on them. They also show the steepness of the land by how close together they are (the closer they are, the steeper the slope).
4) For example, if a map has lots of contour lines on it, it's probably hilly or mountainous. If there are only a few it'll be flat and often low-lying.
5) A spot height is a dot giving the height of a particular place. A trigonometrical point (trig point) is a blue triangle plus a height value. They usually show the highest point in that area (in metres).

A contour map of the Mona Lisa

Sketching Maps — Do it Carefully

1) In the exam, they could give you a map or photograph and tell you to sketch part of it.
2) Make sure you figure out what bit they want you to sketch out, and double check you've got it right. It might be only part of a lake or a wood, or only one of the roads.
3) If you're sketching an OS® map, it's a good idea to copy the grid from the map onto your sketch paper — this helps you to copy the map accurately.
4) Draw your sketch in pencil so you can rub it out if it's wrong.
5) Look at how much time you have and how many marks it's worth to decide how much detail to add.

Q: Draw a labelled sketch of the OS map shown below.

What a relief that's over...

When you're sketching a copy of a map or photo see if you can lay the paper over it — then you can trace it (sneaky). And that my friends is the end of the exam skills section, and the end of the book. Now go treat yourself to an exam.

Index

Index

Index

Index